CON

	Map	viii–ix
	Ammonite	1
1.	Ochre: An Introduction	5
2.	Tin	24
3.	Peat	41
4.	Bronze	67
5.	Silver	94
6.	Radium	127
7.	Aerolite	152
8.	Mercury	184
9.	Copper	213
10.	Gold	246
11.	Lithium: A Coda	277
	Soil	283
	Notes	290
	Acknowledgements	332
	Index	335

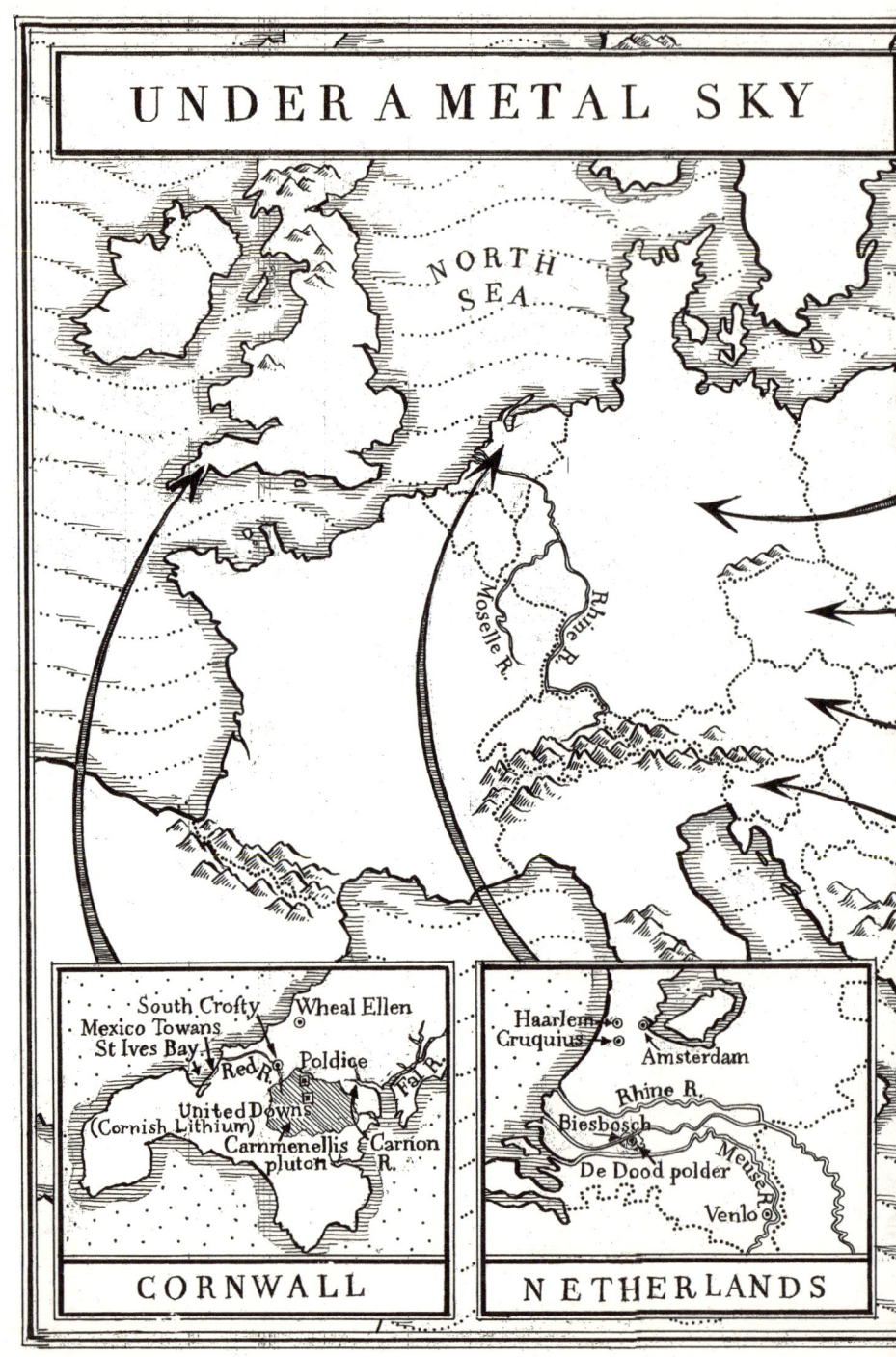

UNDER A METAL SKY

NORTH SEA

Moselle R.

Rhine R.

CORNWALL

South Crofty · Wheal Ellen
Mexico Towans
St Ives Bay
Red R. · Poldice
United Downs
(Cornish Lithium)
Carnmenellis · Carnon
pluton R.

NETHERLANDS

Haarlem
Cruquius · Amsterdam
Rhine R.
Biesbosch
De Dood polder · Meuse R.
Venlo

UNDER A METAL SKY

UNDER A
METAL SKY

A Journey Through Minerals,
Greed and Wonder

Philip Marsden

GRANTA

Granta Publications, 12 Addison Avenue, London W11 4QR
First published in Great Britain by Granta Books, 2025

1 3 5 7 9 10 8 6 4 2

ISBN 978 1 78378 962 7 (hardback)
ISBN 978 1 78378 964 1 (ebook)

Typeset in Arno Pro by Iram Allam
Printed and bound by CPI Group (UK) Ltd, Croydon, CR0 4YY
www.granta.com

For Charlotte

Today, mineralogists recognise that more than 40 per cent of all mineral species on Earth are in some sense biogenic – produced either directly or indirectly through the action of lifeforms.

Marcia Bjornerud, *Turning to Stone: Discovering the Subtle Wisdom of Rocks* (2024)

Höre, du blinder Mensch, du lebest in Gott und Gott ist in dir.

Listen, blind human, you live in God and God is in you.

Jakob Böhme, *Aurora oder Morgenröte im Aufgang* (1612)

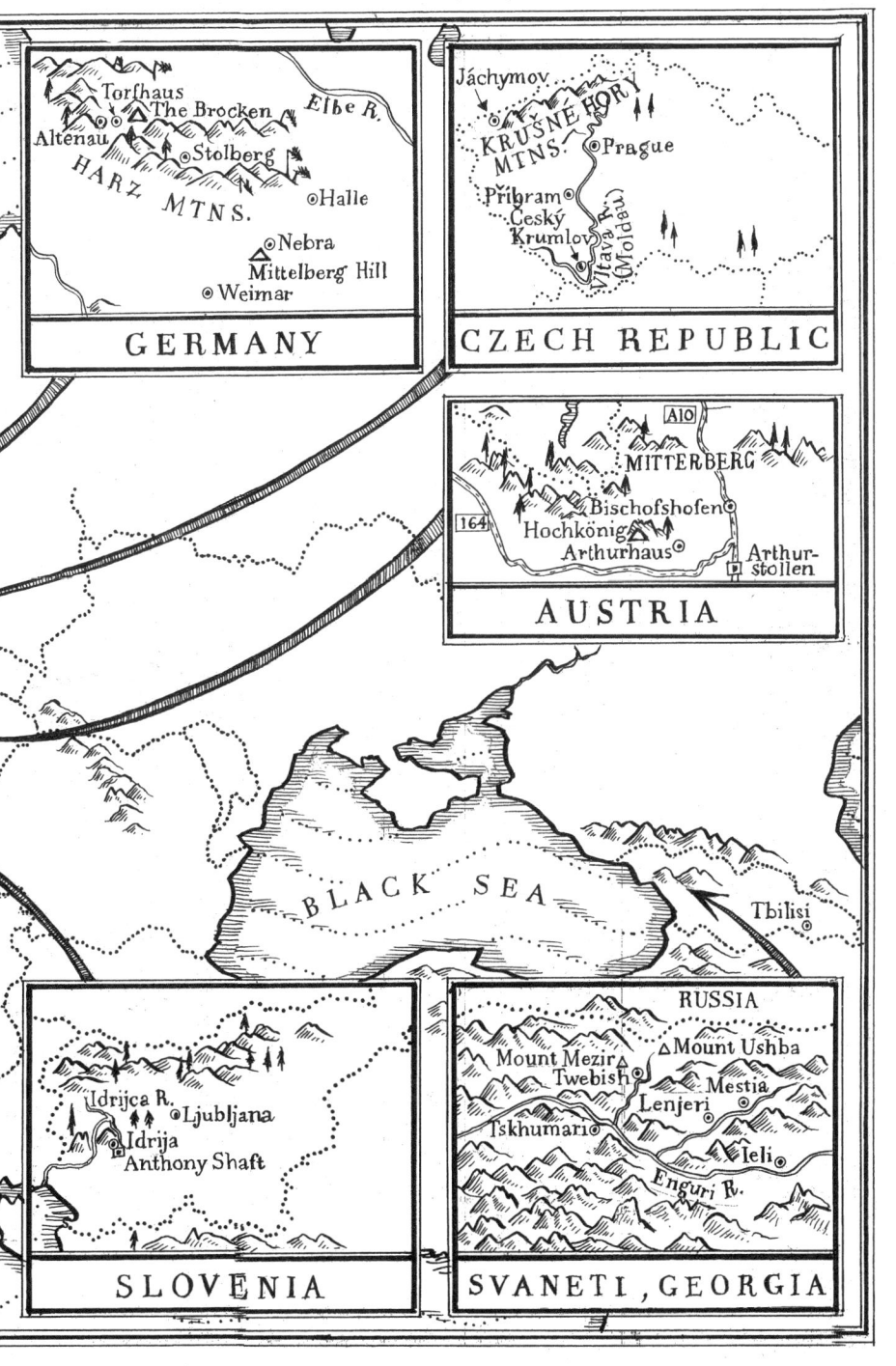

GERMANY

Torfhaus
Altenau
The Brocken
Elbe R.
Stolberg
HARZ MTNS.
Halle
Nebra
Mittelberg Hill
Weimar

CZECH REPUBLIC

Jáchymov
KRUŠNÉ HORY MTNS.
Prague
Příbram
Český Krumlov
Vltava R. (Moldau)

AUSTRIA

A10
MITTERBERG
164
Bischofshofen
Hochkönig
Arthurhaus
Arthur-stollen

BLACK SEA
Tbilisi

SLOVENIA

Idrijca R.
Ljubljana
Idrija
Anthony Shaft

SVANETI, GEORGIA

RUSSIA
Mount Mezir
Twebish
Mount Ushba
Mestia
Lenjeri
Tskhumari
Teli
Enguri R.

AMMONITE

FOR SOME CHILDREN IT'S DINOSAURS, for others trains or mermaids. For me it was rocks. I can pinpoint the moment exactly. I must have been five or six when I heard a surf-like swish, and a truckload of gravel was tipped onto our driveway. I discovered that some of the stones glittered. I put those ones in a tin and sifted the pile for more. I began to look further afield. For Christmas, I was given a geological hammer and chisel and a green canvas bag and went 'chipping' whenever I could. For the next few years, my life became a series of rock hunts, quarry hauls and boulder-flakings, measured out in crystals, fossils and finds.

On a Somerset hillside, I collected 'potato stones', muddy lumps of rock which when broken open revealed sparkling geodes. In Derbyshire, I found a large chunk of Blue John, deep-blue fluorite, which I scrubbed and scrubbed in a youth hostel shower room, and which caught the eye of someone else as it dried, and was gone. Quartz in a dozen shades, tourmaline, jasper, gypsum, agates, gleaming galena, iron pyrites – 'fool's gold'. I scrabbled around on Radstock's slag heaps for fossilised Carboniferous ferns, retrieved serpentine from the Lizard – and from the beach at Charmouth, the cigar shapes of squid-like belemnites.

I didn't think then about the appeal of it, but I remember total absorption, the excitement of the quest and the stones themselves, the other-worldly shapes and colours. Their presence in my room, where I endlessly inspected them, left me with an enduring sense which only later was I able to articulate – that another world lay hidden inside this one.

A couple of weeks shy of my eighth birthday came the big find. I had just started at a boarding school and the alienation of the first weeks was offset by going into the woods with my green canvas bag. One Sunday morning I was out there with a boy called Lea, when he pointed at something that was the shape of a mushroom cap, but much bigger – an old staddle stone. I knew what to do. I took my hammer and gave it a whack. A tiny hairline appeared. Another whack, and we watched the edge of the stone fall away. Exposed were the outer coils of a very large ammonite, a series of perfectly rounded corrugations curling out from the rough sides. It was thicker than my arm.

I can recall now the exact sound the hammer made on the staddle stone, the faintly metallic smell and the shock of its opening. Weeks of chiselling followed, chipping away at the cast in the 'gab-yard', a space behind the school reserved for outdoor hobbies. Groups of boys gathered to look over my shoulders and watch. When fully extracted, the ammonite was half a metre across. Lea and I made an agreement. He had found the stone, I had broken it open. We would each have it for a year and then swap round. He was older than me, and when he left the school, it was his year and he took the ammonite. I didn't see it again.

I carried on rock-hunting, now keener than ever. Grown-ups would ask about my stones. I was the 'geologist'. Aunts and uncles gave me books on geology for Christmas and birthdays. I tried with those books, I really did – I read about crinoids and echinoids, mono-clinal flexures, diagenesis and greywacke. But the truth was that my interest had little to do with geology. I just liked the way these rocks looked. I liked hunting for them, and I liked finding them.

Then came my teens. The green canvas bag and hammer and chisel sat unused in a cupboard. A kind of darkness fell over those years and I ambled through it, going through the motions, keeping the world at arm's length, numb to its real sensations. I look back now and see a lost and aimless figure, like a sleepwalker.

At the age of nineteen, I woke up. I cannot describe it in any other way. It was as if someone had come in and thrown open the curtains and left me blinking in the glare. It happened in an instant. A bus, an urban flyover, an unremarkable evening in October. Out of the window were roofs and more roofs and a big sky and the sun catching the window of a high building. I have no idea what happened in that moment, but from nowhere came a strange sensation. It wasn't mood exactly but it was overwhelming; not a single thought but dozens, tumbling out in a stream which flowed with astonishing clarity, and spread and grew and spread more.

All I could do was look. The same roofs, the same sky, the same buildings. Now they appeared linked and deliberate. Everything appeared linked and deliberate, and every passing notion was coloured with the same ecstatic feeling. Over the next few weeks, it recurred several times so that I had to stop what I was doing and just stand still. It has continued on and off ever since.

I resumed the search, not for rocks and fossils now but for something less solid. Travel and reading bounced off each other. I followed a well-worn path through Buddhism and Taoism and some more esoteric set-ups like theosophy and Rosicrucianism. I grappled with the perennial question – how to square the spiritual realm with the physical? Some years later, writing a book about the Armenians, I came across the Manichaean dualists and the belief that the world was a struggle between the forces of light and darkness. It gave rise

3

to a large number of zealous sects – from Paulicians in Armenia to Cathars and Bogomils in Europe. While many of their followers took negation of the material world to extremes – naked worship, never washing, self-castration – there was something appealing in the Platonic notion of two worlds, of a higher place and the idea that the sensual world gets in the way of it; I remembered my rocks and the impression of breaking through into another realm. Now I think of it differently. The physical and the spiritual are one and the same and they connect, for me at least, through the imagination and the infinite complexity of the natural world.

Recently, I tracked down Lea, my old stone-hunting friend. He was now a wine merchant in London. The first thing he said was: 'When my wife heard you were coming, she told me: "Don't let him take it away – the ammonite was the reason I married you."' He gave an ironic half-smile. 'I'm actually not sure if she was joking or not.'

'Don't worry.' I wasn't sure if *he* was joking. 'I only want to see it.'

The ammonite was propped up on a table in his sitting room. We stood before it, fifty years on, comparing our memories of finding it. They more or less concurred – the stone in the laurel bush, the hammer blow, the surprise of its nut-like opening. I bent to look at it more closely. The pale limestone and its flakiness and texture looked suddenly familiar. They triggered something I found hard to grasp, something more than memory, and I was back in the gab-yard chiselling away, leaning over the rock to tap-tap at its edge, feeling the give as the cast cracked and broke free and another section of the hundred-million-year-old fossil burst into the light.

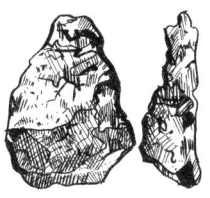

1

OCHRE

An Introduction

OCHRE IS A FERROUS ROCK – clay tinged with oxidised iron, one of the commonest metals on earth. Iron is everywhere. It leaches into soils and run-off. It is held in solution to gather in lake and sea sediments. Pressed down by more sediments, across the galactic span of geological time, it forms mudstone and siltstone – ochre. In situ, ochre can display a spectacular range of tones and patterns – rusty cliff stripes sharp as a blade, or juicy stains that vary so dramatically that exposed rocks look like giant layer cakes.

A very long time ago, far off on the horizon of human history, it was found that you could take a lump of ochre and grind it to a coloured powder. You could mix the powder with liquid – saliva or water or urine – and create a substance that could then be applied to almost any surface. The colour ranged from yellow to orange to brown. It was a pleasing effect and people went some distance to retrieve the rocks, sifting through rubble for just the right tone. They swapped them and traded them, and in this way the colours began to travel further. They were valued. Sometimes the ochre contained flakes of mica or pyrites, adding sparkle to the paint. For that they would trek a little further, ask for a little more in exchange.

Exactly when it began, and how ochre was first used, is not known. A collection of *Homo erectus* bones in Kenya from 285,000 years ago was found alongside five kilograms of gathered ochre. Neanderthals made pigment from ochre as early as 250,000 years ago. At Blombos Cave in South Africa, two 100,000-year-old ochre 'toolkits' were identified: grindstones and hammerstones, antlers, the scapula of a seal – all with traces of ochre on them. Microscopic scrutiny of the traces has captured moments and actions. The ochre grit reveals tiny marks, percussive action and knapping. The stone was sourced from some way away; it was laid on a slab of quartzite and crushed, and the powder placed in abalone shells and stirred, quite gently, to a paste.

From the same site, from the same sea cave come more clues, thirty thousand years later. Now there is evidence of the application of ochre. A stone was found with lines drawn on it – marks on a rock, a fragment. That's all. It took seven years of analysing those marks to work out exactly how they were made, that they were deliberate. The lines are understood to be the first example of 'art', a cross-hatching, an abstract image-making. The marks were not painted on but made by an ochre crayon. Liquid pigment was hard to store and transport, but by adding oil to the powder, you could keep it as a stick, for ease of use. More forethought, more gathering and processing.

At some point, another discovery. If heat is applied to the ochre powder, it changes colour. At around 300 °C, the ochre dehydrates and the iron in it oxidises and changes from yellow to red – the red of raw meat, the red of blood. In their pioneering study of colour and evolutionary linguistics, Brent Berlin and Paul Kay found that the development of colour terms in most languages followed a common sequence – words for black and white were first, but the next was invariably 'red'. The majority of ochre used in prehistory was red.

Ancient ochre use is found wherever there were humans. On the island of Sulawesi forty-five thousand years ago a warty pig was painted on rock, in Australia a kangaroo, in north-eastern Brazil deer and jaguar and capybara, in the European caves at El Castillo, Chauvet, Lascaux and Altamira, bulls and bison and horses – those astonishing forms, all of them depicted in ochre. Iron – in its reds and browns and yellows – somehow conjured to life again the beasts that lived in the imagination as food or threat or both.

Ochre altered forever our relationship with the earth. The dead rock underfoot yielded something miraculous, something striking and powerful, something that with conscious intervention could be transformed, and then be used itself for transformative effect. Some cosmic shift took place in that action – nothing was the same again. Change had always been external, day and night, weather and seasons, rivers and tides, life and death. Now with the use of its own material, the earth could be subtly remade and modified and abstractions created. Dirt was made precious, stones did tricks, rock became transcendent.

Ochre pigments offered little help with the necessities of staying alive – the driving hunger and the cold, or with the making of weapons to hunt with or defend. But they satisfied something for which there was no name, longing of a different kind, a curiosity that would not go away, which justified the labour of collection, the careful processes of preparation, the challenge of effective storage.

Ochre still helps answer questions. Very little evidence of our ancestors survives from the Middle Stone Age – that long flat period before the Late Neolithic and the spectacular leap in the technology graph. Yet anthropologists have prised meaning from traces of pigment production and use. The suggestion is that making the powder does not in itself prove sophisticated thought, nor does creating marks with it. It is the exact nature of those marks that implies an understanding of symmetry and abstract thought. There is also a theory that iron itself – ingested during processing and use – might have helped accelerate the growth in brain function. While powers of expression may have been enhanced by the very material employed for that expression, there were risks too: some ancient pigments contained arsenic and lead and would have been highly toxic. It's a pattern repeated with so many of the earth's resources – benefits and perils going hand in hand.

Colour and dye came first from ochre. As the archaeological record thickens, so the range of decorative stones grows – manganese for black, kaolin for white. In ancient Egypt, blues and greens were highly valued for cosmetics and in tomb paintings; malachite and azurite were two copper-based minerals crushed and turned to paste; quartz crystals were superheated for Egyptian blue faience.

Ochre remained the most widely used of the natural pigments. In Wales, the Red Lady of Paviland (actually a man) is an Upper Palaeolithic skeleton whose bones were covered in red ochre. In Australia, the Aboriginal ochre trade saw pigments travelling hundreds of miles across the continent. In the Mediterranean world, the most highly prized ochre came through the Black Sea port of Sinop, and the ochre itself was known as sinopia. High-quality ochre was also sourced from the Aegean island of Thasos, where people dug into the rock to retrieve it, forming the adits of Europe's earliest mines. Already, on Mexico's Yucatán Peninsula, others had been tunnelling for ochre for thousands of years. Renaissance artists used ochre in frescoes to create initial outlines of faces; it is still used in paints today.

Ochre was the original medium, the first substance to be employed for that strange human faculty – to imagine what is not there, to use two dimensions to recreate the complexities of three, to freeze forever the impression of a single moment, to make solid the ghosts of memory. Ochre elevated all that it was applied to. Every stage of the process had to be followed, from collection to grinding to mixing, but the knowledge once gained was transferrable. It rolled down through the generations, across continents, producing a skill that gave agency to its users and an incentive to hunt further afield, to search for other magical materials that might lie hidden in the earth.

One day a year or two back, a friend suggested surfing a spot called Mexico Towans. I didn't know it; the name intrigued me. It turned out that a couple of young miners emigrated to Mexico in the nineteenth century, then settled there when they returned to Cornwall. Nothing more. *Towan*, on the other hand, is Cornish for 'dune', and the dunes here are vast, part of a four-mile stretch of pale sand that borders St Ives Bay, rising to a hinterland of marram-covered sandhills. The entire seabed of St Ives Bay itself is spread with the same fine sand, many metres deep, and the main source of that sand, washed down over an unimaginable length of time, is two rivers – the River Hayle and the Red River. They drain the uplands of Carnmenellis, Cornwall's main mining area.

It was a beautiful day, mid-October. The sky was cloudless, the autumn winds had paused for breath. Sitting on my board beyond the surf line, I was transfixed by the bay's colour and the swells coming in across it, clean lines that rose from nothing and grew steeper as they approached, and their faces darkened as we paddled for them. Yet what I remember most of that October afternoon is neither the waves nor the surfing but the water itself. The sand below had turned it glassy turquoise. I had the sense of drifting over an alien continent. Down there beyond my bare feet was an endless plain of silica and feldspar grit and in among it were micro-flakes of ores and native metals – tin and copper and zinc, gold and silver.

I'd long harboured a prejudice; I realise that now. In thirty-odd years of living in Cornwall, I'd spent a lot of time in its mining country. It's hard not to. I'd followed paths through landscape scarred by toxic tailings, peered up at the chimneys of old engine houses, through the steel bars of Clwyd caps down into the darkness of long-abandoned shafts. I'd dipped into the books but I'd never gone any further because I thought it would lead only to arid regions of specialised debate, technical details, the finer points of engineering. I was wrong. Over the course of that winter, in a series of day trips and meetings and reading – which led to more trips, more meetings and more reading – I found the subject surprising and expansive. My research deepened as the weeks passed. As I'd discovered years ago, rummaging around among rocks has the strange capacity to open up the whole world. Cornish mining is a story that leads off in some unexpected directions.

The bare facts are well known, but no less astonishing for that. Back in the beginning, Cornish tin helped drive the revolutionary centuries of Europe's Bronze Age. Ingots of Cornish tin, dating from three and a half thousand years ago, have been found in the eastern Mediterranean: present-day Israel and Greece and Turkey. In the early nineteenth century, this tiny peninsula on the edge of Europe was producing half the world's tin and two-thirds of its copper. Of the 450 minerals known to exist on the planet, 14% can be found in Cornwall. Some exist only in Cornwall.

Running down beneath the surface of south-west Britain is one of the world's larger batholiths, where deep areas of magma burst up from the mantle. From the Cornubian batholith came a series of further upwellings. These domes of molten rock, plutons, did not reach the surface but instead solidified several miles below. As they rose, the intense heat separated elements from the country rock and created priceless deposits of metal elements. Fast forward a few hundred million years and the covering layers have been eroded to expose the plutons as granitic uplands – among them Dartmoor, Bodmin Moor and the Isles of Scilly.

Of all these areas, the most heavily mineralised is the one here in central Cornwall – Carnmenellis. Almost circular, no more than ten miles across, the Carnmenellis pluton has always been the most generous of Cornwall's rock complexes. Tin from Carnmenellis travelled thousands of miles from the Bronze Age onwards, as did gold. Its tin and copper helped power the Industrial Revolution. The granite itself has been widely used, kerbing the City of London, rising as lighthouses from the Eddystone and Fastnet Rocks, transported to India and Sri Lanka as building stone. It would be hard to think of another piece of local geology that has had quite such an effect on human history.

I contacted my friend Nick. Now retired, Nick had been Cornwall's County Archaeologist. He was instrumental in the UNESCO designation of the mining area as a World Heritage Site, and before that in saving all the old engine houses from the council's plans to flatten them. There were few mining sites in Cornwall and little about its mining history that Nick hadn't worked on at some point. Like all the best archaeologists he combined rigorous analysis with the understanding that behind it all were living, feeling people. He proposed we meet in the Carnon Valley, up in the heart of the Carnmenellis mining country.

For an hour or two, we pushed up through a network of paths. It was messy ground – centuries of mining had left no part unaffected. Mounds of spoil remained free of vegetation, even after a century or more. We came to a high point and looked out.

'Hard to imagine it all now,' mused Nick. 'Hundreds of people coming and going, and the constant noise. I always think of the noise – stamps thudding and the pumps going and all the horses. Horses and mules everywhere! At the height of it all, there were more engine houses working in this little valley than in the rest of the world combined.'

Unseen beneath our feet ran defunct levels and stopes and adits. The old shafts were capped to stop people and animals falling in. The opposite slope was full of cramped plots of pasture, the traces of miners' smallholdings. Not everyone did well from mining, or even managed to support themselves.

'What I always think about these places,' Nick appeared suddenly animated, 'was that it was *heroic*, really heroic. I mean, I know it was grim and it probably didn't feel like it at the time, that it was all about money and reward, and there was certainly nothing altruistic in it. But something about the scale of the work, the guys just hammering away for hours, and the inventions they had to make. Just staggering what they did.'

It is often said that the spectacular trajectory of our species was set in place by the Neolithic Revolution, the transition from hunter-gathering to farming, planting crops and animal husbandry. Agriculture was born, and with it the notion that we can stand in the flow of the universe, stand still and divert it for our own ends. Yet extracting and processing the earth's resources has had just as great an impact.

It started with ochre, continued with copper and tin and iron, and went on through the periodic table. Rocks that were just rocks in one era became essential stuff for the next. The ground revealed itself as a treasure chest, a vast store of miraculous substances, and metals were the real prize. Ploughshares and knives, money and nails, lunulae and torcs, axes and mattocks, hair clips, belt buckles, armour and goblets, ships' hulls and aircraft fuselages, electric cables and semiconductors, nanochips, superconductive micro-materials, lithium-ion batteries. Metals were buried alongside priests and queens in ancient tombs, and now orbit the planet as satellites and lie along the ocean beds as plaits of pulsing messages and data. Metals have made us heroes, given us the wings of birds, the speed of the wind, the voices of gods.

All things have their price. Naysayers pointed to the dangers early on. Nature kept its materials underground on purpose. Cut into the flesh of the planet and you wound it. Use its resources and suffer the consequences. In *Metamorphoses*, Book I, Ovid condemns the work of miners:

> they dug up riches, those incentives to vice, which the earth had hidden and had removed to the Stygian shades. Then destructive iron came forth, and gold, more destructive than iron; then war came forth.

Metals turned man against man. They promoted inequality, rewarded greed. They lethalised disagreement. They set up a compelling new set of priorities that offered immediate gain and long-term wealth. The rocks from mines and quarries elevated people, or some at least, created dynasties of emperors and kings, formed armies and filled state coffers. But there was a cost. 'Works of silver and purple are of use not for human life,' lamented Socrates, 'but for Tragedians.'

In Greek mythology, Prometheus stole not just fire but the art of processing metals. It was he who encouraged men to do more digging for

> The other boons,
> Stored in the womb of earth, in aid of men—
> Copper and iron, silver, gold withal—

And there he stands, chained to a rock in the Caucasus while every night his liver is pecked out by a giant eagle. The gods don't take kindly to mortals pilfering their powers.

Stone, Bronze, Iron. In the nineteenth century, the Danish archaeologist Christian Jürgensen Thomsen introduced the categorisation of prehistoric ages that is still in use today. Each age was a simple technological step forward. Such a progression made sense to those in industrial Europe who saw improvement all around them: bronze was better than stone and iron was better than bronze.

In one of the earliest surviving poems of the Western canon, Hesiod proposed something similar – Golden, Silver and Bronze Ages, then a Heroic Age and an Iron Age. Hesiod's system was not based on materials but on moral quality. For him, there was no progression, no improvement. Quite the opposite. Each one was a diminishing, a further step away from the original Golden Age. He himself was living in the Iron Age, and he didn't like it.

'I wish,' he proclaimed, 'that I were not any part of the fifth generation of men but had died before it came.'

There's a pattern to it all, a narrative arc worthy of Greek myth or tragedy. First comes wonder and benefit, then power over others, and finally over nature itself. Only later do the full consequences become apparent and the realisation that something else had come up from the ground when the ores were dug out, and that was hubris.

The wind was freshening, driving hard from the east. In its gusts was the chill of the coming night. Nick said there was an old processing site he wanted to show me. We walked on up, further into mine-scarred landscape. Below us in the sump of the valley lay a reservoir. The water was not for local use. In fact, it didn't look like water at all. Even in the gloaming the colours were a fierce palette of sickly green and orange – metallic, unnatural, repellent. The dam had been built in the 1980s as a tailings lagoon and had filled with a greyish sludge. Water had dissolved the iron in the sludge to produce a deep ochreous tinge, as if a giant carcass was rotting. In the upper edges of the lagoon, the rust-tinged water shallowed to an algal fringe the colour of luminescent mould.

Skirting round it, we came to Poldice. At the heart of Carnmenellis, Poldice mine had once been abuzz with activity. In the early nineteenth century, it employed up to a thousand people. A local rhyme went: 'At Poldice the men are like mice'. The parish – Gwennap – was known as 'the richest square mile to be found anywhere on earth'. Then in the mid-nineteenth century, tin and copper prices collapsed. The machinery fell silent. Hundreds of workers – men and women – were laid off overnight.

Carnmenellis, though, had another card up its sleeve, another mineral whose value had recently been increasing and was abundant in the area. A new processing plant was built, for arsenic. In the cold dusk, we reached its collection of roofless ruins: granite-block furnaces and a network of ancillary buildings. We stood among them.

'This is the calciner,' explained Nick. 'The ore was heated on plates, and the fumes collected. They were channelled into a labyrinth – 'lambreth' in Cornish – which cooled the fumes. The arsenic crystallised on the walls and then the workers came in and scraped it off.'

Arsenic had long been useful. Like ochre, it was a pigment and dye in prehistory, producing beautiful shades of green and yellow. It was a depilatory for tanners, an additive to strengthen bronze. It had health benefits – it was taken as an aphrodisiac or general tonic; for ulcers, lice, shortness of breath; and as an expectorant. Hippocrates recommended it as an escharotic. In Chinese medicine it was used for fevers, to clear obstructions and as an elixir for long life.

Ayurvedic texts cite it as a laxative, a blood purifier, a treatment for asthma, bronchitis, leprosy and fear of ghosts. By the nineteenth century, it was being widely administered in Western medicine. Skin diseases, blood diseases, rheumatic diseases, malaria, diabetes, diphtheria, tuberculosis, cancers, syphilis, scrofula, snake bites – all could be treated with arsenic. 'At no other time in human history,' wrote the medical historian Jerome Nriagu, 'has the health of nations depended so much on one element.'

When nature placed arsenic in the ground, it played a trick. It was the same trick concealed within so many of the earth's materials, from lead to gold, radium to fossil fuels. Use just so much and no more. Arsenic was both magical and malign, practical and poisonous. Colourless, tasteless and odourless, it is the toxin of choice for murderers. The dose can be regulated for a precisely timed death – sudden, gradual or lingering. Its effects on the body are hard to trace. Nero used it to kill his brother-in-law. The Borgias laid down stocks of arsenic-based *cantarella* for the removal of cardinals, kings and popes. In seventeenth-century Rome, the contract poisoner Giulia Tofana used arsenic-laced cosmetics to kill hundreds of people, before being discovered and executed.

The early Chinese used arsenic as a weapon – 'holy smokes' were capable of poisoning large numbers of soldiers at once. In 1858, twenty-one people died in Bradford after a batch of peppermint humbugs was inadvertently cooked up with arsenic. Among wealthy Victorians, wallpapers became fashionable, particularly ones with brightly coloured arsenic pigments. Rumours of poisoning led Queen Victoria to order all green wallpaper to be stripped from her residences; at the same time, a number of children in London died from arsenic-tinted wallpaper. The artist William Morris had produced some of the most popular designs and dismissed the idea of toxicity as 'witch fever'. He was a family shareholder and a director of the Devon Great Consols mine, which was the world's largest producer of arsenic.

Nick and I scrambled over the ruins, into the labyrinth itself – a series of narrow switchback corridors. Clouds were visible overhead, above the open-top walls, skidding across a darkening sky. The men who'd worked in here were careful to wear mouth coverings and to stuff their noses with cotton wool. There was no record of serious poisoning but it was said that arsenic gave the Poldice workforce particularly fine complexions and silky hair.

Much of the arsenic was shipped across the Atlantic. It proved very effective against the Colorado beetle, and potato farmers were encouraged not to hold back in its use:

> Spray the scale that's hiding there,
> Give the insects all a share;
> Let your fruit be smooth and bright,
> Spray, O spray and do it right!

From the late nineteenth century, about four hundred million kilograms of arsenic was deployed on crops in the US. It is estimated to stay in the soil for about nine thousand years.

The spoil-heaps at Poldice are still free of plants. In the half-darkness I squatted down on one of them. There was a crust to the surface, which crumbled when I scratched at it with a rock – a mixture of fine gravel and silt. Something was missing. It had no softness and no smell. Its sterile texture was devoid of soil, that fertile matrix of a billion microbes, all feeding off the decay of organic matter.

Arsenic exists everywhere in small quantities, and in small quantities it promotes health. Extract it, concentrate it, and it's a killer. In the sixteenth century, the alchemist and physician Paracelsus was one of the first to identify the paradox: 'All things are poison, and nothing is without poison; the dose alone makes it so a thing is not a poison.'

The Forge and the Crucible is Mircea Eliade's classic study of metallurgy in traditional society. In it he proposes an astonishing idea – that 'the *imaginary world* (*univers imaginaire*) . . . came into being through the discovery of metals'. The production of metals, he suggests, was more than a functional process. It was revelation. It permeated all aspects of social life – belief, economics, war, sexual practice. And it stemmed from the impact of a single moment, the miracle in a furnace, when shining metals first trickle from dull rock. Witnessing such a transformation, Eliade believes, altered everything: 'Not only did the manipulation of metals contribute considerably to man's conquest of the material world; it also changed his world of meaning. The metals opened for him a new mythological and religious universe.'

It is impossible now to know how much metals actually 'opened' the imagination, but they certainly led it in a particular direction. Among people with animistic beliefs, the earth was already a living thing. Metallurgy elaborated the idea and gave rise to the widely held notion that the soil was a womb-like space in which substances gestated. Base metals like lead and iron were engaged in a series of alterations towards perfection, towards gold. That in itself encouraged the quest to discover and replicate those processes. Over time the techniques grew more intricate. For some the quest was borne of reverence, for others of greed. In Europe it became known by its Arabic name, *al-kimiya*, or alchemy.

One of the most striking things about alchemy is how widespread it was – more so than any one religion. There was alchemy in early China, Ayurvedic India, ancient Egypt and Abbasid Baghdad. It was harnessed by kings and emperors, who employed teams of highly trained workers to try and distil the inner workings of the cosmos. Some of the greatest alchemists came from the Arab world; many were Persian. Alchemy re-emerged in Renaissance Europe, spreading around its capitals and courts – before being eclipsed by the glare of reason and the Enlightenment.

Alchemy was long seen as the bachelor uncle of true science, a rakish figure with impulsive and indulgent tastes, but without issue. As a subject for serious study, the history of alchemy was considered untouchable. Academics are no longer so wary. In its arcane texts they are finding, amidst the mumbo jumbo, things of value. Where alchemy was once dismissed as pseudo-science, it is now considered proto-science, credited with a good deal of the methodology and spirit of the Scientific Revolution. Alchemists always tended towards being figures of fun, but in their deluded idealism remains a view of nature that is more and more relevant for our own troubled age. They took an epiphanic moment – the emergence of metal – and built from it a belief system, an entire discipline of applied enquiry, driven by devotion and awe and the conviction that the world around us is a place of miracles waiting to be uncovered.

I went further afield, up to Bodmin Moor's mining areas, out to the far west around Land's End where the old levels push out beneath the seabed. I spoke to retired miners, who recalled the terrible conditions and, in the same breath, how much they missed it all. One of them mentioned a group known as the Beardy Weirdies, who met once a month in a social club near Redruth. I looked them up on Facebook: *Beardies is for ex-miners, mining enthusiasts, underground explorers, geologists, mineral collectors. Please bring artefacts.*

On a damp evening, I pushed open the door on a hubbub of pub chatter, with a scattering of beards but also many clean chins, all nodding and drinking and talking, figures mingling around tables of rocks and old miners' lamps and general clobber. I spoke to a group of elderly miners, a rock collector from out of county and a man who produced an ingot he'd made himself from alluvial tin. When he turned it over, the underside sparkled. 'See that? Gold. I won't tell you where I found it – but it was in Cornwall.'

I also met Trevor. In a cinnamon-coloured jersey and heavy neck-chain, he was sitting at a table with a pint and an impressive collection of large crystals. He told a remarkable story. Originally a Londoner, he'd been on a bus in Peckham when he spotted an advert: *Retrain! Miners wanted in Cornwall.* He didn't know where Cornwall was but because he was 'in a bad place', he set off west to apply. He soon found himself underground. He liked the miners and he liked the work. 'Look at the size of me! I was an electrician in London – small spaces are not a problem.'

What he remembered most about his mining was nothing to do with tin. It was a chance find. 'One day we set the charges at the face and when we came back, I noticed a small hole up in the ceiling. It was a "vug", like a void in the rock. I climbed up into it.' He found himself standing in a space entirely filled with crystals. 'Practically blinded me!' Soon afterwards the mine closed, and he left Cornwall to work on the Channel Tunnel. But he never forget that vug. Years later he came back. Although the mine was now inactive, he managed to re-enter it and found the geode. It was damaged but over the course of several weeks, he extracted what crystals he could in a wheelbarrow. 'Now they're bloody everywhere in my house.'

For days afterwards, I found that image returning – a figure standing inside a crystal chamber. It would have sent my old stone-hunting self into a frenzy just to think about it. I thought of it somehow as a picture of yearning, of the highest hopes we place in the earth. Rocks and metals are pursued not just as resources but to satisfy that strange abstract hunger that follows us all. During those months, I became aware once again of their revelatory appeal, convinced that an immersion in them might yield something about the big questions – about desire and greed, reward and risk, gifts and poisons, the way we see nature and the shifting stance from which we view and treat the material world.

Just below my studio, a short walk across a field, run the last tidal reaches of the Upper Fal. It is now a deserted creek, an oak-fringed backwater clogged by centuries of sediment. Yet several thousand years ago, during the Bronze Age, the valley was navigable many miles upstream, to not far south of Goss Moor – and around Goss Moor were found some of the highest concentrations of alluvial tin in Europe.

Once I knew that, I began to see the valley in a different light. I imagined small craft manoeuvring down between the tree-lined slopes, tin ingots lining their bottom boards. I pictured those ingots downstream being transferred to larger ships for onward passage – from the entrance to the Fal deep into Europe and down to the Mediterranean to be chemically wedded with copper to make bronze. It was a coupling with profound consequences.

I pieced together the outline of a journey – beginning here in Cornwall, with tin. I'd follow these Bronze Age ingots eastward into Europe, up the Rhine to the heart of the continent. I assembled a host of stories and half-stories, tales of rock enthusiasts, alchemists, mystics, ancient mines, calamities and wonders. They were the bare bones only, but like any mine prospector, like my 'chipping' expeditions of old, it was a case of identifying a likely area then setting off in hope to see what was there, to see what I'd find.

The story of the earth's metals and minerals is by no means only a European one – the skills of metallurgy are thought to have developed independently in many unconnected places: in the Balkans, the Caucasus, northern China, Egypt, West Africa, North and South America. But something happened to Europeans when they discovered the materials beneath their feet. It stirred an energy that was part curiosity and part avarice, a restlessness that, in due course, brought an end to the Middle Ages and sent fleets of ships off around the world in search of more materials, more resources.

Out here on the far western tip of the continent was a good place to start, out here where the Variscan orogeny left in its wake an intense concentration of useful and potent metals – and none more so than tin.

2
TIN

TIN IS ONE OF THE rarer metals. At two parts per million in the earth's crust, it is thirty or forty times less common than nickel or zinc or copper. It is soft and easily cut or bent. Just before snapping, as the crystals align, it gives out a little whimper. Like most metals, tin has its own quiver of super-properties. Its low melting point makes it an important solder. Flat glass with astonishing clarity is now produced by floating it in a bath of molten tin. Because it does not readily oxidise, tin has been widely used as a food preservative. In 1812, two British merchants bought from a Frenchman a patent for food tins and built the world's first cannery in London.

These properties have a flipside. Below 13 °C, malleable alpha-tin becomes the allotrope beta-tin, a transformation that turns it brittle. That same year, 1812, saw the collapse of Napoleon's Russian campaign and the retreat of half a million French soldiers from Moscow. Harried by Cossacks, weakened by hunger and cold, the infantry soon became aware that the tin buttons of their uniforms were crumbling, allowing the wind to drive through their tunics and into their bones. Thousands froze to death. Now tin's association with war is as a 'conflict mineral', its extraction, along with that of other metals like tantalum, cobalt, gold and tungsten, helping to fuel local militias.

In recent years an ultrathin material has been developed out of tin which is just an atom thick. Known as stanene or '2D tin', it is billed as a 'super-material' that might even surpass graphene in the effect it will have on our lives. Electrical current can pass through stanene at room temperature with 100% efficiency, no resistance and no heat, pushing superconductivity to another level, allowing the scaling-up of quantum computing and a further miniaturising of technology.

The origins of tin have always been something of a puzzle. It is not entirely clear how it is deposited in lodes nor how its ore is formed, nor its precise relationship with the granite with which it is always found. In the ancient world, when bronze was the metal of choice, the copper was easy to source (the island of Cyprus gave the Romans the word *cuprum*, 'copper') but only very small quantities of tin were available. Archaeometric analysis has now revealed Cornwall and south-west Britain as the main source of Europe's Bronze Age tin. It had to travel far, and long-distance trade was one of the era's many innovations. It led to the rich mixing of peoples and ideas, songs and stories – the great ferment from which modern Europe emerged.

The word 'tin' has its root in Germanic languages. In Greek it is '*kassiteros*' and in Latin '*stannum*'. The Cornish '*sten*' comes from the Latin – pointing to the importance of Mediterranean markets; in Welsh, its sister language, the word is '*tun*'. Herodotus was the first to mention the Cassiterides or 'tin islands' – 'whence we get our tin'. The name has been thought to refer variously to St Michael's Mount, the Isles of Scilly or Britain as a whole.

It was long supposed that the Phoenicians dominated the Atlantic tin trade, that they made use of the alluvial tin deposits which were abundant in south-west Britain. Amidst the speculative tangle of prehistory and etymology, the connection to the Phoenicians became swollen in importance: they brought to Cornwall clotted cream, pasties, darker skin, a tin figure found in Bodwen and a bronze bull from near St Just-in-Penwith; the name 'Britain' came from *bre-tin*, with *bre* being Cornish for 'hill', and the 'hill of tin' clearly being St Michael's Mount. The Cornish themselves were said to have Phoenician blood. It's all nonsense; quite possibly the Phoenicians never even reached Cornwall.

In the early nineteenth century, a fishing boat in the Fal estuary dredged up an ingot of ancient tin. The St Mawes ingot was H-shaped, the four arms enabling it, some say, to be strapped to either side of a mule or horse. It was thought to have been on its way from upriver, and from there it would have continued out into the English Channel, down into the Bay of Biscay, up the Gironde River, then on to pack animals to take it across to the Mediterranean. For many years it was on display in the Royal Cornwall Museum. I was helping there one day when it needed to be rehoused. Its size gave no idea of its weight; it took three of us and a trolley to move it, heaving it inch by inch across the floor. As we struggled, I had the sense that this sea-pocked lump of metal was hiding something inside it.

It was at a meeting of the Beardies that I first met Sam. He was standing in a ring of people discussing whether a particular place deep in Wheal Jane was a prospecting drift or a winze; it was clearly a matter of some importance. Sam was a mine explorer, part of a group who spent their spare time crawling around Cornwall's subterranean network of adits and stopes and shafts. If a dog fell down an old mine – a common occurrence – they were the ones the police called on. Later we spoke at length, and he said: 'We're going sluicing this weekend. Come and join us.'

We met in the dune country, in a sandy car park just along the coast from Mexico Towans. Sam was in the cab of his friend's truck when I arrived, his back against the door, and he was laughing. 'Pat here was just telling a story. He was saying how he once . . . *Really?* Actually—' Sam opened the door and jumped down from the cab, 'you don't want to know.'

Sam hadn't explained what sluicing was exactly. As we unloaded the gear, I could see it involved a twelve-volt bilge pump, a car battery and a long galvanised trough – a home-made sluice lined with the green tangle of plastic matting called miners moss. It also involved a storm stove, buns, sausages, a sledgehammer, two iron stakes, ropes, helmets and several clinking sets of carabiners and harnesses. There was a perfectly good path to the beach, but we were going down the cliff. To Sam, I soon learnt, an expedition that did not involve ropes wasn't worth going on.

His father had worked at South Crofty, Cornwall's last tin mine. He died when Sam was just ten. But it wasn't his father that led to his interest in Cornish mines and metals. It was a piece of land that he bought near the village of Nancledra, where his family had originally come from. 'I kept some goats there and chickens, until my English-bastard neighbours made it all hell – and I sold up. On the land there was this old waterwheel – huge bloody thing in cast iron and timber – and it once ran a tin stamp. Got me wondering about the whole process. And, well – once you start looking into mines, you don't stop.'

Sam hadn't stopped at all. He had read and researched everything he could find on Cornish mines. He had his own small archive and collection of minerals and mining paraphernalia. His conversation, jokey and ribald for the most part, was also informed by years of fact-finding. His love for the mines was his life.

We ferried gear to the cliff. A number of others were already there – among them Cakey Teaboy, so called because he liked to bring a cake ('My advice?' said Sam. 'Do not eat it'), and Jenn, who ran ultra-marathons when not running St Ives Bakery; she'd also brought a cake. A few of them abseiled down to the beach, and before following them, the rest of us lowered the equipment on the same ropes. One of the bags broke free and bounced and thudded into the sand, leaving a trail of ketchup and mayo on the cliff.

'Fucking amateurs!' shouted Sam. 'The lot of you!'

The site itself was a small cut-back in the cliff where a stream of water leaked from a pipe and flowed out across the sand. Beneath the pipe was a slick of ferrous orange. We dug out the sand where the stream crossed it and placed the bilge pump in the pool that formed. The pipe fed water up to a rocker box on top of the sluice, and it was already flowing down through it. Then the work began – digging sand and dumping it in the sluice. The mounds sat for a moment before flopping from below, collapsing into the stream, pouring through the trough, which was set up at a particular gradient.

'About 20% for tin,' said Sam. '8% for gold.'

'You sluice for gold?'

'Every Cornish river's got gold in it.'

From time to time the sluicing stopped and the miners moss was rinsed out and the sludge collected in a large bowl. Then the digging resumed. Sam panned the bowl, swirling the water to sift off the lighter sand. We took it in turns to dig with two Cornish shovels and also a conventional spade or, as Sam put it, 'a piece of English shit'.

In that way the morning passed. By the time it took to cook and eat the sausages, we had two bowls of sludge. Further sifting and panning had purified it, but still the sand was only about 5% tin.

'What'll you do with it?' I asked Sam.

'Take it to my friend down near Breage. He'll run it through a centrifuge, put it on his shaky table. We'll then fire it and make an ingot.'

'An ingot?'

'When I've added in the stuff I got the other day, and do a bit more, should be about like this.' He held open his thumb and forefinger by about ten centimetres.

I thought of all the effort to get here, all our digging and sluicing. 'That's quite a small ingot.'

He looked at me with mock outrage. 'But it's *tin*!'

Like gold and silver, tin touches something deep in the core of our being. 'This darling Metal,' wrote William Pryce in the eighteenth century, which 'among the working Tinners holds her Empire in the heart'. In his monumental *Mineralogia Cornubiensis*, Pryce described the extraordinary lengths Cornish miners go to retrieve it: 'unremitting ardour, patience, industry and resolution scarcely parallel in any other . . . undertaking under the sun'.

Such ardour did not mean an easy life. In Cornish mines, accidents were frequent and often fatal. The daily grind of working underground, in poor air, constantly damp, the drilling and hammering, the driving of charges into the lode, shinnying up and down ladders, took its toll. In later years, when there were figures for such things, over half of the tinners died with silicosis.

There are few first-hand accounts of early Cornish tin-mining. Miners didn't write, and anyone who could write didn't want to go down a mine. One exception was James Forbes, who in 1794 was on a tour of the West Country and felt curious enough to don the bug-ridden protective clothes and descend. On all fours he crawled through a stream, then dropped by rope 'down a chasm, no broader than a chimney'. There, at the heart of the whole operation, he 'saw two figures that hardly wore the appearance of human beings' working at the tin lode. What surprised Forbes was not so much the cramped and hellish space – but that these men 'were *singing at their work* [his italics]'.

The pursuit of tin was always enigmatic. It was far removed from the normal food-producing round of plough and scythe, seed and stock. When they first appear in historical records in the late twelfth and early thirteenth centuries, Cornish tinners had long been occupying the margins. The 1201 Charter of the Stannaries embraced age-old exemptions and privileges, stating in law that tinners were different, that they should be treated differently, 'as by ancient usage they have been wont to do'. Most taxes did not apply to them. They did not have to pay tithes to the Church nor do military service. They were allowed to dig in any unenclosed land. They had licence to cut turf and divert streams. They had their own legal system, their own courts and their own gaol.

Working in remote country, tinners also developed their own dialect, a muddy combination of Cornish and English. A 'dippa' was a 'pit', a tin-rich stream a 'beu-heyl'. The slime left by the tin was 'loobs'. A sieve for sifting sands came in many forms – it was either a 'dilluer', a 'ridar', a 'kazer' or a 'searge'; to shake it was to 'toas'. Close working of the tin lodes, the object of the entire enterprise, bred more terms. Near the surface you might encounter the 'bryle' of the lode, already partly broken up and in it might be a harder stone or 'cal'. A route to the main lode could be via a 'string' or 'guide', which would need 'dizzuing', exposing on just one side. A large lode was a 'squat', and a piece of dead rock within it was a 'horse'.

To look for tin was to step into another world, one populated by its own deities, who were hypersensitive to any intrusion of human will. Naming animals was one way to provoke them. So a tinner would never talk of a 'fox' but instead say 'long tayle'; likewise, a hare was a 'long ear', a cat a 'rookes', an owl a 'long farcer', while a rat was a 'peep'. Anyone using the received name was dispatched to the alehouse to fetch a gallon of beer.

The area worked was often barren, poorly soiled or boggy. But to the tinners it was a living place – a skin of alluvial deposits rich with metal, or surface lodes to pursue down into the body of the rock. The names of the setts reflected a colourful animation of place, each one a story in itself: Piskey Meadow, Turn About, Cost is Lost, Sweep-All, Best-to-Agree, Space Bean, Little Dribble, Fat Work, Be Lovely, Wheal Tesgentle, Wheal Saturday, Two Pair Goodlucks, Shillings-go-by, Great Groan, Ployden's Misfortune and Rusty Hammer.

A week later, I had a call from Sam. He was going down Wheal Ellen. 'There's a bit of new ground I want to look at.'

The old shaft at Wheal Ellen lay in a scrubby piece of land, toxic and scarred from long-ago mining. It had been capped. A protective breeze-block square had been built around the shaft, about chest-high, and covered with a lattice of steel bars. The bars were rust-brown and one of them had been cut at some point, leaving enough room to squeeze through. The other bars provided anchor points. Sam was starting to set up.

Squatting on top of the breeze-block wall, I looked down through the bars. The shaft was squared off at the top; one or two timbers had survived from the original shoring, but mainly it was just bare soil and polypody fern and moss, which thinned as the level fell and the edges faded into black, into darkness.

Sam was fixing a second belay. 'Fancy it?'

'Why not?' Studied nonchalance, inner anxiety.

We were four now. Jenn had turned up. And Dave, a window cleaner with an amused half-smile. When I asked what appealed to him about mine exploring, he said. 'I can't stand heights.'

'But you're a window cleaner?'

'Oh, that's OK. It's down there I can't stick. Fucking hate these shafts.' Confronting his vertigo when he pulled on his harness had cured him of a long-term depression.

With the rope secured, we let it fall down the shaft, coils flicking from the pile as it ran out of the bag. When it stopped, Sam said: 'Forty-two metres, give or take.'

Wheal Ellen is not the biggest, nor the deepest, nor the oldest. It's just one of dozens of mine complexes around the Carnmenellis granite dome, and many more around Cornwall. The number of old mines is unknown, incalculable. Some were little more than holes, speculative probings after a lode that soon thinned or disappeared. Others were more enduring, and the cassiterite ran deeper, opening out to leave behind a vast wormed-out tangle of winzes and levels; in many places they joined up as one team broke through on another. Often there were fights when that happened, but once they were resolved the old Cornish 'Wheal' names were replaced with 'United' or 'Consolidated' or 'Consols'. Many were boarded over when they closed, 'sollared', but soil gathered on the boards, hiding the trap below, and then the boards rotted. Now there are frequent cases of livestock falling down old shafts. Sometimes home-owners wake to find that their front lawns have turned into great whale gullets in the night.

When Sam first mentioned the idea of 'dropping' Wheal Ellen, I looked it up on the catalogue of Kresen Kernow, Cornwall's national archive, housed in an old brewery in Redruth. I may not know about setting up the rig to drop a mine, but I knew my way round an archive. I found an early map of the mine and called it up – a roll of card with meticulous penwork following its subterranean routes. Each level was a different shade, and there were spot depths and inset sections for each of the shafts. It was a work of art and I carefully took pictures on my phone.

I showed the photos now to Sam. 'I thought these might be useful.'

He took a quick glance. 'Here Dave, you slag, fetch us that strap.'

I put the pictures away.

Inside the shaft-top were a couple of iron steps set into breeze blocks. Clipping a cow's-tail line to the bars overhead, I climbed down the steps and stood to fix the main rope to the descender. Sam talked me through it. Unclipping the cow's-tail, I let my weight settle in the harness. I squeezed the handle, the rope ran through it, and I dropped.

The air changed almost at once. It became musty, spore-thick to start with and then just heavy. My light picked out the sides of the shaft as they slipped upwards – the broken rock and then the solid bedrock. I thought of the initial digging, the days and days spent swinging picks, raising spoil in buckets. Mines like this were not one-off projects but worked and re-worked over generations.

Penduluming slightly, I held out my feet to stop bumping the sides. Stones and grit scuffed loose and the sound faded as they fell – BOUNCE . . . bounce . . . thud.

Down and down. All I could see now was what my light picked out: the cone of layered stone. High above was a shrinking dot of sky.

'You OK?' Sam's voice was muffled and distant.

'All good!'

I was fine. I felt secure in the harness. The underground didn't bother me. I'd grown up in limestone country and had done a little potholing. The only thing that did worry me was the fragility of it all – the idea that these old works, their wooden props and back-filled material, might be a little precarious after a century or so abandoned.

My boots brushed ground, fumbled, and gripped. I stepped out of the harness. While the others came down, I ducked under an arch and into the mine.

It was a brown world. Soft brown underfoot and hard brown of the rock. There was a little grey mud and some yellow, but they too looked like shades of brown. The black on the roof was concentrations of iron, but when you looked closer you could see they were dark brown. The quest for the pure shine of tin had left little of colour behind.

Ahead were several openings. I walked for a little way along one and found a chute leading from it. I looked up and could see that the backfilling, and one large boulder, was held up by a single, rotting prop. I returned to the shaft.

Sam had appeared now with the others. 'You saw that rock? Each time I come down, I expect it to have fallen. Everything's on the move down here.'

I remembered a definition I'd read somewhere of these abandoned places: *Cornish mines – collapsing structures of toxic rocks.*

'OK, all ready? Let's get going. Be careful not to touch any of them timbers. That includes you, Dave.'

'You don't have to tell me, mate.'

I could see now why he was so dismissive of my photos. They'd had the precision of an architect's drawing – a schema to present to backers and investors, who'd never go underground. The reality was far less tidy.

We pressed on, down the tunnel, passing other side-workings. There were larger chambers and galleries. In places the path was open on one side, dropping away into darkness where the lode had been worked out, down to the next level and beyond. And sometimes the path itself was cut away and the empty space filled with boards, now soft and weakened, and you made sure to walk down one side, the side with solid rock below, one foot squarely in front of the other.

Then the passage came to a stop. At our feet the ground dropped to nothing. A few metres on, it resumed. A length of softwood trunk lay propped over the darkness. Links of rusted chain hung from it, down into the void.

'What we must do,' said Sam, looking around, 'is, er . . . take a couple of firm steps – hand on the wall here. And Dave? Don't look down.'

'Yeah, right.'

Dave did as suggested, and was over. I put one foot on the log.

'What's your blood group?' quipped Sam.

'Ha-ha.' A few quick steps, and I reached the other side.

Jenn followed – nimble, assured. 'Phew!'

We dropped down a level, clambering over a big brown boulder, and on through a collapsed section where rock and mud filled the way ahead. A squeeze and a crawl became a prostrate wriggle over a collapsed section – elbows and knees, inch by inch, with metallic dust in the mouth – and a drop at the end, so that you had to lower yourself on your hands, and half-tumble as if coming out of a high cupboard.

In the rubble I spotted a rusty shape. A chisel. It was bloated out of shape by rust-oysters but somewhere inside was the blade and the mallet head. I placed it back. Elsewhere we came across other signs of work abandoned: an old wooden barrow, a broken earthenware flask, the stem of a clay pipe, candle stubs, the felt rim of a miner's hat. At one point, where two paths met, was a small slab of clay stuck on the rock. It had rows of holes in it, like a cribbage board.

'Awesome!' Sam shook his head. 'Tally board, I reckon – they'd use it to count barrows, maybe – or men who'd gone down that way...'

There was still a peg in it.

Mines like this were vacated overnight. A price drop, a loan called in, and the payroll was closed, the engine turned off and soon sold. Now they exist as a suspended moment, frozen in the state of the final day. Water and gravity continued their slow ministry, oxidisation its transformings. But the frenetic industry ceased in an instant, its note sustained faintly in these traces, here many fathoms below ground.

Sam was leading when he put out a hand. We halted. In the mud ahead was a faint heel-sized crescent – and stud marks, several in fact, regularly spaced – the hobnailed imprints of a miner's boots. Then we really did feel the ghosts.

'Oh, Christ!' Sam gave a shake of his shoulders. 'I got tingles down my back.'

A tinner from a century and a half ago – whistling perhaps, familiar with this place in the dim light of his candle, not knowing it was his last day. Sam urged us to walk along the rock to the side, to leave the mud intact, and we all shared his reverence.

Down another level, and now there was more water. The adit was calf-deep. We sloshed through it and the tunnel amplified the sound. The water was the colour of Heinz tomato soup. At some point, maybe last winter, it had flooded and all the iron in the rock had turned to rust. Orange sludge covered the walls.

Sam and Dave stopped to investigate a side-route. Jenn and I pushed on, down the adit. The orange mud thinned and the bare yellowy brown rock was exposed. Suddenly we were among stalactites. They started as thin finger-like bars pushing down from the ceiling. They were brown, of course – and glistening. The acidic water was full of iron hydroxides, and left a tiny deposit as it dripped. On the floor were corresponding studs, upturned bulbs of burnt brown. We stepped more carefully, crouching to avoid knocking them off with our helmets. Then there were more. These ones were larger – thick as a thigh at the top, or extending to drapery formations which reached from ceiling to floor, and we had to slalom to avoid them. We stopped, found ourselves chuckling, touching them.

'They're unbelievable!' called Jenn.

'Beautiful . . . in a completely weird way.'

They were smooth to the touch, rippled and rounded by constant wet. What was striking was not just the size of them, nor their sculpted form, nor the number of them – but their speed. Calcium carbonate stalactites in limestone caves grow at about 0.13 mm per year. The largest, in show caves – with their organ-pipe formations – need tens of thousands of years to take shape. These ones, these post-mining shapes, had formed in less than two centuries. Elsewhere in the mine we had seen blue copper growths – not as big as these – but already having created flowstone shapes in the same short period. The blue itself was extraordinary, covering the rock in sheets, and Sam had said: 'If the miners'd seen that, they'd have chased it back to the lode.'

Just as the miners themselves had dug out these imitation cave systems in super-quick time, so the copper and iron had decorated them with rapid speleothems. Perhaps that is the real superpower of metals – the hidden accelerator in their chemistry, revealed to those who first shaped them, who learnt to smelt them, to forge ploughs and spears. Slow labours were made quick. The saunter of time became a trot, then a gallop. Metals offer a fast route to change, the sort of change that is normally undetectable – rivers forming valleys, mountains growing or species evolving. It wasn't tin that was being sought here, nor the money it might bring. It was the power of immediate transformation. And since the last miner left, when easier and cheaper sources were found in South America or Australia, the unextracted metals have continued their time-lapse progress unseen, staining the walls with their improbable blue or transforming the empty tunnels into colonnades of oxidised iron.

When we joined up with the others, they were standing over a pool of water. It was so still and so clear that it looked less like water than a faintly altered pocket of air. The colours were polished, with a bluish tinge in the shadows. And in among it all were billowing shapes of algae, weightless as clouds so that after a while staring at them, it seemed we were not looking into a pool but at some distant skyscape. We walked on, astonished.

Sam was excited because on the way down he thought he might have found another entrance. We started to climb back. At a certain junction, with a mine explorer's hound-like sense, he had detected the air was a little lighter and fresher. Now we crawled through a half-blocked winze into a long gallery. At the far end was a stumble of rubble, a shoulder-height climb to make before entering another, smaller gallery.

'Can you feel it now?' called Sam.

There was a faint change. The mustiness mixed with clarity. A slit exit from the gallery meant a crouch and then the passage closed and dropped, and we were down to hands and knees. The air was definitely different – and there was a smell too. Hard to identify what exactly – something between an old attic and the bin area of a hotel. There was a muddy section to wriggle through. Pushing with my elbows, I followed Sam's boots, and watched them move clear.

'Oh— my god! *Fu-uck*! Take a look!'

I elbowed out and scrabbled to my feet. In the roof was an opening – an old shaft. But the opening was no longer open. From it bulged a huge mass of – what? It looked like the cod-end of a vast trawl, frozen in the moment it dropped the catch on deck. This was not fish. It was rubbish – a blackish-brownish mush of degraded waste. There were flecks of white plastic in it, and strips of blue. Looking more closely, I could see yoghurt pots, bleach tubs, old feed bags, fertiliser sacks. A suitcase, its material gone, could be made out by its skeleton frame, and that too had begun to crumble and rust. The bulk of it all was a dark oily mass, where moisture had washed down the decaying material from above, dissolving it to essence, an essence of trash.

We walked around it.

'I'll tell you what it is,' Sam nodded to himself. 'Before there was rubbish collection, farmers used these as dumps. I've seen shafts with a bunch of old cars down them.'

'Yeah,' recalled Dave. 'Like a vertical bloody traffic jam.'

We poked around and looked at the packaging. Sam picked up a shred of yellow newsprint. 'The seventies, at a guess.'

None of this was meant to be seen again. Shoved down here – out of sight and out of mind. It was certainly never meant to be seen like this, from below. I had the feeling that we'd entered the throat of some dead beast and were looking up at its last semi-digested meal. There was something sickly about it.

In one corner I stumbled on a small bunch of plastic flowers. They were generic flowers with red and blue petals and green leaves. The years had done nothing to wither their shape or colour. I pictured this place in hundreds of years' time, when the rest of the waste had been reduced to a brown omni-sludge – and these flowers still here, still firm of stem and still in bloom.

It was early spring. The hazels and hawthorn were in bud. The afternoons were inching forward in daily increments of light. From my studio, I walked down towards the Fal. Through the trees I could see the foam-fronted tide creeping across the mud. Unseen behind it, molluscs were opening up to feed, while greenshank beak-probed in search of them. The winter birds were still here – lapwing rising from the far side of the creek, flashing in the sunlight, and plover and godwit. In the field a couple of short-eared owls had been around for months. In a week or two, they'd be gone, off to the wetlands of Scandinavia and the tundra of the north.

I too would soon be leaving, following the course of Bronze Age tin as it left this valley and headed up the English Channel to enter Europe by its mouth. In the territory where the Rhine, Meuse and Scheldt rivers merge was a vast swampy region that has since become the Netherlands. It yielded a resource that was extracted with the zeal usually applied to metals, and which had many of the same unintended results.

3

PEAT

PEAT IS ONE OF THE earth's great gifts – the multi-use gift
of charged black earth. In its anaerobic layers, decomposition is
slowed and organic material retains its energy. Dry peat, and it
burns. It gives steady heat. Cooking on peat is slow and even, good
for casseroles and stews; pans are easier to balance on the cut turfs
than on logs. You can place potatoes straight into the hot ash, likewise fish or eggs. In the peat-rich lands of Ireland and Scotland, it
was considered the life force of the house. Each night householders
would mutter prayers to the goddess Brighid – 'Brighid of the peat-
heap' – as they banked up the peat-ash to keep the fire going until
morning. In the 1960s, 40% of Ireland's electricity was produced
from burning peat.

As well as energy, peat gives metal. Watercourses, heavy with
dissolved iron, flow into peat bogs and stall and stagnate, and
the metal is drawn from the water and made solid by acidity and
hordes of bog bacteria. Peel back the top layer of peat and you'll
find pea-sized pieces of iron. Collect them, and a few years later you
can visit the same piece of bog and find new nuggets. The spread
of Viking settlements – from Norway to Newfoundland – can be
traced by proximity to peat bogs and bog iron. Rich in silicates,

the hard lumps produce a shiny metal which resists rust. Viking longships – which helped shape the history of medieval Europe – derived their strength from nails and rivets of bog iron.

In the Finnish *Kalevala*, the smith discovers bog iron in the marshes, in the paw prints of passing animals:

> Then the blacksmith, Ilmarinen,
> Thus addressed the sleeping iron:
> Thou most useful of the metals,
> Thou art sleeping in the marshes,
> Thou art hid in low conditions,
> Where the wolf treads in the swamp-lands,
> Where the bear sleeps in the thickets.

And as well as energy and bog iron, peat also gives health. Its peculiar biochemistry means it can remove wrinkles, treat psoriasis and herpes, and act on tumours and viruses. Therapeutic peat baths became hugely popular in the nineteenth century, and sphagnol soap was used to treat facial wounds and trench sores. As a sorbent, peat can draw harmful dyes from effluent or extract heavy metals from wastewater. When the tanker *Full City* ran aground on the Norwegian coast in 2009, peat moss was used to mop up globs of black fuel from water and rock.

And as well as all these gifts, peat also reveals the past. Seamus Heaney, who called peat 'kind, black butter', pointed to the many exhibits in the National Museum of Ireland which bear the caption 'found in a bog'. And sometimes the find itself was butter – at more than five hundred sites in Ireland have been found deposits of bog butter. The earliest are from the Bronze Age, and no-one is sure whether the butter was deposited as a votive offering or to improve the flavour.

In Ireland and Scotland peat has always meant survival. Its vital importance remains embedded in the language, a lexical conglomeration which, like bog iron, is replenished through use. The word in Scottish Gaelic for a quaking bog is *sùil-chruthaich*, literally 'the eye of creation'. In Irish, peat is not just peat but is distinguished by type. *Cinceach* is 'brittle peat' and *scileach* is 'brittle black peat' and *clochmhóin* is 'hard black peat' and *méithmhóin* is 'brown peat'. A rick of drying peat is a *stualainn*, or maybe a *gróigeán* or a *dúchán*, and what was left of last year's turf is *athmhóin*, and every variation in rick or type of peat or cutting-tool has a name, because for centuries the extraction of turf was life itself.

Peat's anaerobic matrix removes time from the process of decay, and in its spongy mass are retained long-vanished beasts – mammoths and lions, bison and wolves. Humans too – Haraldskær Woman and Cashel Man and Yde Girl. You can gaze into the face of Tollund Man or Lindow Man (aka Pete Marsh) and have the fleeting sense that you know them. These bodies generally reveal a passive and violent end – garrotted, strangled, stabbed, clubbed, before being buried in the peat. Were they too offered like butter to propitiate the bog's own gifts?

Barely covering 3% of the earth's current land surface, peatlands hold twice as much carbon as all the world's forests combined. But when you use peat, when you dry it or cut it or drain it, the carbon is released into the atmosphere, where it traps the sun's heat. In the last fifty years, one hundred thousand square kilometres of peat bog has been lost.

The biggest peat bogs in the world are in Siberia. One alone is the size of Germany and France combined. They've existed for thousands of years. They are not too dry and not too wet, not too cold and not too warm, and their populations of tiny liverworts and mosses thrive, and when they die the peat retains their carbon

43

dioxide. But temperatures in these regions are rising far faster than elsewhere. The frozen parts of the bogs are melting. The peat is being flooded in places and dried in others; either way, carbon storage is breaking down and every summer now there are fires. Once ignited they are hard to extinguish. Stamp out the flames on the surface and they go underground, smouldering in the dried peat, burning unchecked for years. These 'zombie fires' are releasing more and more of the forty billion tonnes of carbon held in the peat – and that means more icecap melting, higher sea levels.

Peat and rising waters are a combination that is well understood in the Netherlands. For many years, the Dutch enjoyed the gifts of peat. In the early modern period, while neighbouring countries were held back by wood shortages, peat helped the Dutch lowland-ers to surge ahead. The Netherlands was the first country to draw political power from thermal energy. After 1600, Dutch industry developed rapidly, driven by easily accessible peat – Delft pottery, ships, building materials, bread, beer and sugar (half of Europe's sugar refineries were Dutch by 1662). This small boggy country then spread its presence across the world's oceans, set up lucrative colonies in America and the Far East and created an artistic and intellectual milieu that can be ranked alongside Italy's Renaissance.

But each turf taken opened up the soft soil. Ponds became lakes. Then the lakes joined up to form one large, peat-fringed sea in the centre of the country. The Haarlemmermeer kept on growing, and in the 1830s storms drove its waters right up to the edge of Amster-dam and no amount of windmills could stop it getting bigger.

J ust to the south of the Dutch city of Haarlem, in the village of
Cruquius, I stood before a castle. It was not a big castle but with
its neo-Gothic style, its battlements, keep and a forbidding-looking
moat, it displayed a certain flamboyance. In the middle of the nineteenth century, it was built around a weapon that had been brought
in to defend the Dutch against their most persistent enemy. The
weapon was a beam engine and consisted of a single, 144-inch-
diameter cylinder. It was the largest steam engine ever built. When
the windmills failed, the pump was installed, and with two other
similar pumps it worked continuously for three years to drain the
Haarlemmermeer.

I'd approached the Cruquius pumphouse after a long walk
from Haarlem, following a neat pavement which bordered a neat
cycleway and a neat four-lane main road, from which other roads
led neatly off. Low cloud smothered the flat land. Cars purred past
in the mist, muffled and regularly spaced. Everything at the bottom
of this old lake was so orderly, so tidy in its bicycle-thick, grid-
housing-block modernity that it was easy to forget that the entire
place is below sea level, that vigilance and dikes are all that keep you
from flailing about in five metres of water.

The Netherlands and the Rhine delta have long been the entrepôt
to northern Europe. In the Bronze Age, it was one route that tin took
from south-west Britain on the way to its liaison with copper. Some
three thousand years later, the Cruquius pump followed a similar
path. The pump was constructed in Cornwall, on the northern edge
of the Carnmenellis mining area, and it was the work of Harvey's

of Hayle, who had become experts in pumping water up from mines. Nothing, though, that they'd encountered underground quite matched the challenge of dealing with the Haarlemmermeer.

The story of the Netherlands and its peat bogs is the story of our wider relationship with the world around us. It began with an unpeopled wilderness. During the Roman period, the land around the Rhine delta was no place for the living. Near-permanent fogs fused sky and earth, while the swampy flats themselves were an ambiguous interzone – neither land nor sea but a dangerous combination of the two. Set foot there and your life was at risk. Water levels were constantly shifting. What was solid could turn liquid at any moment.

Then a hardy people came from Sweden and settled in the far north, on the Frisian Islands. They lived a marginal existence. They built long dwellings on little atolls that they formed themselves from dragged clay. In such forsaken land, any allegiance was irrelevant. They paid neither tax nor tithe. When they died, according to Procopius of Byzantium, they were exempt even from paying the ferryman of the dead. Some believed they *were* the ferrymen. Pliny the Elder was stationed nearby and reported that these poor people 'try to warm their frozen bowels by burning mud, dug with their hands out of the earth and dried to some extent in the wind more than in the sun, which one hardly ever sees.'

By the ninth century CE, other plucky settlers arrived. The peat landscape, they discovered, was not entirely flat but formed shallow domes often many miles wide. Though the top of those domes might be several metres above sea level, the sphagnum sucked up the water and ensured that even at the top of the mounds the surface was always wet. By cutting radial ditches the settlers managed to drain the domes and make them habitable. They placed churches on their summits. They raised dwellings and byres. They set fire to the dried-out ground to concentrate nutrients and minerals. The land became not just habitable but lush, and in summer was covered in green pasture and happy cattle and crops.

The peat itself proved key to living here. Drained, it became fertile; dried, it was fuel. Soon word spread inland that there was a cheap alternative to burning wood. A nascent peat industry emerged. In the west, towards the sea, inside the line of high protective dunes, the peat was burnt, and from the ashes came salt for the herring industry.

What had once been a soggy hell became a place of plenty. Food and warmth and mineral health emerged from the mire. The years passed. Isolated farms grew into villages, and villages grew into towns and cities. By the time of the Reformation, the low-lying provinces of Holland and Zeeland were among the most prosperous and densely populated regions of all Europe.

The whole enterprise was based on a single piece of technology. The *klepduiker* is a non-return valve, a flap that opened when water pushed against it, but closed if the flow was in the opposite direction. In this way the early Dutch managed to get round one of nature's fundamental rules, stopping water from going where it wanted to go. The challenge of living at sea level was solved. The *klepduiker* helped drain the swamps and created in their stead large areas of fertile land. The Netherlands flourished.

Trouble, though, was on its way. Nature might wait a little, but in the end it will come calling, knocking on your door to claim its debts. If you don't pay, it will break down your door. Which is what happened. Because peat is predominantly water, draining causes it to sink. Once dry, the long-prevented process of oxidisation begins. The soil broke down, causing further sinking. Often the land dropped by several metres. There was now no slope for the water to run away. In many cases the gradient was reversed and the dried ground became wet again, the water seeping into arable soils. 'Grain does not like wet feet,' muttered the farmers.

Thousands of people left the land and migrated to the rapidly growing cities of Haarlem and Amsterdam. Fortunately, there was plenty of work making bricks and dyeing cloth and brewing beer and distilling gin and, fortunately, there was also an endless supply of fuel for all those industries and new homes. Boats lined up at the quays to unload their peat.

Behind them, back out in the marsh, something was happening. When the peat was cut and the peat-cutters had taken it away, water began to seep into the emptied-out hollows. It dissolved the edges. Small ponds formed. When the wind rose, it made little waves, and the peat started to crumble. The ponds became lakes and the lakes joined up. The land was disappearing and many people found it hard to believe that it could be linked in any way with the cities' new prosperity, with all the fuel-hungry kilns and furnaces. They imagined something evil at work in the fens, a mythical beast. They spoke of *de Waterwolf*.

Catastrophe, when it came, came swiftly and it came often. Storms in the North Sea, tidal surges, dune breaches, river risings – water had any number of ways of breaking into these sunken and exploited peatlands. The most robust of dikes could collapse in hours. From the twelfth century onwards, a series of inundations devastated the land. Entire villages were wiped out; thousands died.

Each time it happened, rather than retreat, the people applied the twisted logic of survival. They built higher dikes, mud embankments that enclosed sub-sea-level pockets of ground – polders. The great poet of the Dutch Golden Age, Joost van den Vondel, wrote a poem in which the Dutch lion does battle with the Waterwolf. The peat-farmer – *de veen boer* – looks on, cheering as the lion gains the upper hand.

All human communities exist by overcoming natural constraints – the Dutch more than any. 'God created the world but the Dutch made the Netherlands,' goes the popular saying. What began with gravity and drainage progressed to engineering – to sluices and locks and sea walls and larger dikes, to the great circular channel of the Ringvaart, and finally to moving water upwards, using wind. The first windmill to pump water was built in 1408. Soon they were everywhere.

But windmills could not prevent the great creeping disaster that took shape in the middle of the country. By the sixteenth century, the four lakes had joined forces to form the Haarlemmermeer. A battle was fought on its waters in 1573 between a fleet loyal to the Spanish and the piratical Sea Beggars. As the inland sea doubled in size, then doubled again, calls to drain it became louder. Throughout the eighteenth century, schemes were put forward – mainly involving more windmills. None of them succeeded. In the 1840s, the Haarlemmermeer's waters appeared on the cities' streets. The Dutch government set up a commission: How to drain the encroaching waters? More windmills, agreed the commission. But the king had seen the power of steam and he'd heard it was used to good effect in the far south-west of Britain. He sent two men to the mining districts of Cornwall.

In this way Harvey's of Hayle, which a generation or so before had been forging shoes for workhorses, received the royal contract for three steam engines. Harvey's had to turn to other forges, other engineers around Cornwall, to fulfil the order. The Cruquius pump was the last of the three in service. It worked until 1932. At its closing-down ceremony, the officials professed the widespread love felt for the machine, what it had done for the Dutch people, what it had done for their country. They placed a wreath of flowers on its copper cap, as if it were a fallen warrior.

The pump has recently been restored. Every Monday now, the *Stoembruck* or 'steam team' come in to maintain the pump. Inside the castle, I found one of them stretched out on the floor beneath the machine. Heaving himself out, he pulled a rag from the back pocket of his boiler suit. 'Giving the old girl a little grease.'

He was Australian. He'd been born in the Netherlands not long after the war, but he'd left when still a young boy. 'Later on I went backpacking and like every bloody Aussie, I ended up in Earl's Court and like every bloody Aussie, I went and fell in love!' She was Dutch and they ended up living here, not far from where he was born. 'So, full circle.'

He went to the wall and pressed some buttons. A thud, a clank, some remote whirring, and the cap began to rise. It paused at the top, then dropped. Above it, the huge beams pivoted. Out of the window, I could see the bloom of water in the pool below. The pump is now driven by electricity, but during its working life tonnes of coal had been shovelled into its boilers, one fossil fuel deployed to fight the damage done by another.

The pumping house is now a museum. Out of the window and beyond the pool, there was real work going on. The bucket of a large digger – diesel-powered – was tamping down a slope of new soil on the bank of the Ringvaart, the original canal that still runs around the dried-out lake of the Haarlemmermeer.

'Repairs?' I asked.

'Always. That's one damn job that will never end.'

In Amsterdam, I met up with a windmill expert called Jippe Kreuning. An article he'd written – 'The Many Faces of the Waterwolf' – had come up in an online search, and we'd struck up a correspondence. Now we were sitting in a canal-side bar roaming around a number of subjects – traditional boats (each lovingly showing pictures of those we sailed on, like they were family), archaeobotany (in which Jippe had a PhD – examining the contents of medieval Dutch cesspits and their surprising amount of Middle Eastern spices) and windmills (he had visited Iran to look at the world's oldest, but was not convinced of any direct link between them and the first Dutch ones). We spoke about the Waterwolf and the pumping of the Haarlemmermeer, and I then asked him a question that had been on my mind: 'With rising sea levels, hasn't this country more to worry about than most?'

'You'd think so. Many people appear to believe the opposite. We are the ones who have always managed to deal with water. People have this idea that we'll cope.' He shrugged. 'Who knows if they're right?'

It was Jippe who first told me of the Biesbosch, the 'forest of rushes', a small piece of abandoned polder down in the south of the country. Reclaimed in the Middle Ages, then flooded, then reclaimed again and finally abandoned to nature, the Biesbosch stands as a monument to the country's relationship with peat and water. Not far south of Rotterdam, Europe's busiest port, the Biesbosch is now uninhabited, a national park, a wild and swampy area of some thirty-five square miles amidst the labyrinth of waterways around the Rhine–Meuse–Scheldt delta.

Jippe gave me the name of a local historian. Wim van Wijk lived just outside the Biesbosch. 'You'll find he knows plenty about the place. He's written seven books about it.'

I phoned Wim in the morning and he mentioned the 'Biokeet', a hut in the heart of the Biesbosch that was used by scientists who came to monitor the area. 'I think it is empty now. I can arrange for you to stay there.'

He said he'd send me a map, and a link to a canoe-hire company.

That afternoon, I took a train and the last bus to the town of Hank. The bus driver gave a demob grin as we both got out. 'No bus tomorrow – King's Day! We are all drinking.'

One of the patterns of travel, in my experience, is that whenever you pitch up in a new place, it'll be a public holiday. The next morning, groups in orange gathered in Hank's town square – orange T-shirts and orange skirts, dayglo orange hats and orange-rimmed sunglasses, and Dutch tricolours crayoned on the cheeks of the children. People were setting out stalls of bric-a-brac. 'Ach!' sneered one elderly lady. 'Look at them selling their old rubbish no-one wants.'

I headed out to the Vissershang marina on the edge of the Biesbosch and there picked up a canoe and loaded my supplies. It was a Canadian-style canoe, basic hire-quality for knockabout afternoons and family picnics. I had little knowledge of canoes but had found a YouTube video kindly supplied by 'Total Outdoorsman', in which he explains how to paddle solo. To prevent the bow swinging off, you apply an 'outward pry' at the end of each stroke. 'Simple as that,' says Total Outdoorsman. I tried it as I left the pontoon. The bow swung round. Practice, practice.

The pontoons and masts slipped astern. Ahead was a winding corridor of high poplars and waterside willows. They enclosed the sounds – the gentle splash of my paddle, the fat plop of a carp and the sudden squawk of an Egyptian goose. Three days here! My excitement was tinged at its edges by the doubt that remote places always bring – is this *allowed*? – and an awareness of what might go wrong, particularly when boats are involved.

In the Rijksmuseum hangs a diptych called the *St Elizabeth's Day Flood*. It shows this area, the area around the Biesbosch, in the early fifteenth century. The left-hand panel celebrates rural plenty and fertility – villages built on mounds with ornate church towers and large barns, plump cattle and carts full of grain sacks heading for the city. The polders here had long been the breadbasket for the great medieval port of Dordrecht.

In the second panel is the same impression of a rich agricultural region, the same plump beasts, the same grain sacks. But in the top-right corner, something is happening. Water is pouring through a broken dike. The animals, seen more closely, are hurrying away. The human figures are heaving grain and paddling barges, making for the shelter of the city. One or two bodies already float in the rising waters.

The event itself was sudden. On the night of 18 November 1421, the North Sea surged and breached the dikes. Dozens of villages were swept away, several thousand people died. Centuries of prosperity were wiped out within hours. There were other floods in the region – successive incursions that made the land untillable again, the pasture ungrazable, the villages uninhabitable. An estimated one hundred thousand people lost their lives in the five great Dutch floods between the twelfth and fifteenth centuries. But in collective memory, the image that stands out most is the *Elizabethsvloed*, the St Elizabeth's Day Flood.

And the cause? River surges, political tensions, lapsed dike management – all those things, but cited again and again: 'peat-mining'. On the seaward side of the dikes, free-for-all digging had been taking place for centuries – for fuel and for salt. Many attempts were made to stop it but the commercial imperative always won out. After the flood, Dordrecht – the oldest city of Holland and once its capital – was an island. It went into a slow decline. The polder that fed it became a watery wasteland, testament to the perils of extraction.

Slowly the waters shallowed. Silt was deposited by the rivers and reeds began to grow and people came by barge to harvest them. Then more silt came and the reeds gave way to rushes and the area became known as the Biesbosch – 'the Forest of Rushes'. And the rushes were harvested and the land dried further. Then willows were planted and pollarded and the withies sold.

The Biesbosch is now a crucible for a new method of land management – that is, no management. It has been left to nature, a version of how the story began, and an overflow for increasing river floods. Human activity has given way to beavers and wildfowl. There is talk of introducing moose.

'Head for De Dood polder,' Wim had said on the phone. 'You will find the hut there. But I must apologise, Philip.'

'Why?'

'The name – it means "death". No-one's really sure why. Probably someone died there. Next door to it is Moordplaat – and that means "murder place", so probably someone was murdered there.'

For several hours I paddled west and north, through a series of canals fringed with thick reeds and thick trees and birdsong – with an abundance of life, in fact. Wim's map more or less conformed to what was here. I took side-routes if they looked interesting, enjoying it all too much to worry where they led. This was roadless, watery country, ruled by flying things – from the single white-tailed eagle to the spoonbill and egrets on the wing, to the blue flash of a king-fisher and a nesting coot plump on its reed throne, down to slender damselflies and a red admiral – and for King's Day, deep down one of the side-canals, fluttering across the bows, an orange-tip.

By mid-afternoon, my happy meandering had turned into simply being lost. I found myself in a semi-blocked waterway. Ducking beneath collapsed willows, I paddled on. The vegetation thickened, the open water closed. Out of stubborn principle, I did not turn back. When I found myself dragging the canoe over a partially submerged bough – balancing on it to pivot the boat over – I gave in and retreated. Doubling back, picking up the right direction again. Paddle, paddle, *an outward pry* ... It was working better now. The banks slid by, tangles of root and reed and bramble, the uncut hedge-tops of once-pollarded willows. It had the feel of somewhere less uninhabited than abandoned. I pressed on, towards the polders of Death and Murder.

De Dood turned out to be a very pleasant piece of raised ground with a sea of rushes beyond it; and the Biokeet was not the bleach-board wooden hut I'd pictured but a sturdy brick building, with basic wooden cots upstairs, a kitchen with an old sofa, and a field lab with lots of galvanised work-surface. I stowed my food, rolled out my sleeping bag, then went for a walk.

It was late afternoon. I picked up a path out to the east, padding along the top of an old, daisy-topped dike. The ground was soft underfoot, dapple-shadowed from overhead poplars. To one side was a dried-out creek and on the other, traces of willow planted in lines.

I came to an old hawthorn bush. I'd half walked past it when I stopped and looked again. There's something so perfect about hawthorn when it's healthy and strong and in its spring garb. And the smell! Thick and sweet ... Some people don't like it – it contains trimethylamine, the first chemical to be released by rotting corpses and also present in sexual secretions. Life and death, *de dood*. This one was a crescendo of white blossom – it was hard to see any green on it at all. I stood before it, marvelling, trying to identify the complicated notes of its scent – sex, death? It smelt more like cake.

Soon the land opened out and I found myself on the edge of mudflats which stretched out to the south. Who's here? A binocular scan – creatures familiar from Cornish creeks: shelduck, Canada geese, six redshank, spindle-legged and nodding as they pierced the soft silt for food. Still here, or perhaps on passage. Sometimes there is a much rarer spotted redshank among them – but not now. The shapes on the mud were lapwing, now rising, peeling off the ground with their lonely *pee-wit*. A posse of terns came in overhead, turning on slender wings, quick-gliding and banking, speed without effort. And another group, slower and stouter, with a final flurry before landing and the coarse sound as they foraged – black-headed gulls, their faces dipped in tar.

I sat on the side of the bank, idly watching the birds. An hour passed. I saw the colour of the mud change, and the colour of the sky. I read for a while. I followed an oil beetle struggling through the grass at my elbow. Bank slope, mudflats, the distant backdrop of trees, and all of them giving fixity to the to-ing and fro-ing of beast and bird. Yet few places in Europe are less fixed than here, few places give the lie better to the permanence of the shoreline. The layers beneath me were historical, their deposition measured in hundreds and thousands of years, the recent era of humans, not the millions of years of geological time.

About fifteen millennia ago, when the last ice retreated and people began to spread into northern Europe, all this was high above sea level. The coastline was far to the north, stretching across what is now open sea between England and Denmark. With the retreat of the Devensian ice, there came a rapid and sustained rise in the water. When I first saw the graph, I didn't believe it. The rate picked up from about 10,000 BCE. It peaked at a rise of thirteen metres over a two-thousand-year period. Thirteen metres! That was about twenty centimetres during a thirty-year lifetime. An astonishing figure – particularly when it was sustained, and particularly when the land – Doggerland – was pretty flat. Large areas would have been lost every generation. During this time too, in about 6200 BCE, came the Storegga Slide, a glacial slipping: the sea dropped suddenly by twenty metres, then came a twenty-metre tsunami that pushed the sea tens of miles inland in a matter of minutes.

Doggerland was well used by Mesolithic people. It offered good hunting for fowl and fish. There would have been middens and seasonal settlements and pathways. But the cumulative experience, during those centuries and millennia, was a diminishing, a backing-away from an ever-retreating shore. Firm land turned to swamp in a generation, with the aquifers suddenly brackish, graves and ritual sites inundated, an abiding sense of loss, of physical ambiguity as solid ground turned to water, and rapid change became the norm.

In about the fourth millennium BCE, just as communities began to farm and clear forest and plant crops – the start of a radical altering of the world – the rate of sea-level rise slowed. Silt from the Rhine–Meuse was pushed up into high dunes which still form the Dutch coast. Peat built up inside the dunes, a thick stagnant sponge of organic material. As people settled these boggy wastes, they drained the peat, extracted and burnt it. The era of floods began.

The region has known no serious floods for a couple of generations. But floods are never forgotten and exist for us all in the deep-down strata of cultural memory – the biblical flood, the Gilgamesh flood, the Gun-Yu flood of China and perhaps, deeper still, the worldwide inundation that drowned Doggerland. It is estimated that by the year 2100, the global rate of sea-level rise will approach something similar to the fastest pace of the early Holocene.

The wind was picking up. I could hear it in the poplars. The leaves made a watery sound; their white undersides flashed like ticker tape. There was an edge of winter still in the wind and I felt suddenly ill at ease, alone in this place with darkness coming. I rose to leave – and as I did so, I spotted a roe deer on the edge of the reeds. Its head was twisted towards me. Even at a distance I could sense its muscles taut, its frame quivering. Fear and enchantment filled the space between us, and then it turned and was gone.

Draining swamps is one of the supreme expressions of human will. It's always been popular with autocrats and dictators who seem to think that swamps are put there to defy their vast and wonderful power. In the eighteenth century, Peter the Great dried the wetlands at the mouth of the Neva river in order to embellish the earth with the classical splendour of St Petersburg. It was a high calling, to banish the bog and give his people the capital they deserved, to prove themselves truly European. Tens of thousands of serfs died in its making; in 1824, floods drowned hundreds of the new city's residents.

'Whoever drains swamps,' boomed Frederick the Great of Prussia at the head of his mighty army, 'is making conquests from barbarism.' His chosen water-purging was achieved with the graft of his soldiers. Dike-building and river-straightening, they laboured to liberate the flood plain of the Oderbruch from its unruly damp. As he surveyed the vast expanse of reclaimed land, Frederick summed up his achievement: 'Here I have conquered a province in peace.' Thousands of wolves were slaughtered, along with boar, bear, lynx and – according to one source – upwards of eleven million sparrows; in 1997, 13,500 people were forced to flee floods in the Oderbruch.

More recently, Mussolini drained the Pontine Marshes and installed two thousand farmers to prove his success. Saddam Hussein drained his southern marshes to rid them of the rebellious

Marsh Arabs. Donald Trump's 'draining the swamp of Washington' was rhetorical, but it drew on the idea that swamps are in need of draining, and that such intervention, by its very nature, leads always to improvement.

It doesn't. Goethe's *Faust* is a tragic fable of European progress and its dramatic climax is a scene of swamp-draining. Goethe worked on the play for sixty-five years and although famously hard to stage, it manages to pull together the deep undercurrents of his time and to create something with enduring resonance. The final pages – Faust's death and the fate of his soul – were completed just before Goethe's own death.

Faust's life has been a long struggle for meaning, a quest to find some lasting fulfilment on this earth. One day he looks out at the sea and is struck by its great pointlessness:

> It flows along, rising to inundate
> The lowland bogs, brackish and desolate.
> The waves come rolling in again and again
> And then recede – and where, tell me, is the gain?

Faust knows at once what he must do: 'I'll bar / the lordly sea out from the lowly shore!' He'll banish water from the swamp.

Faust is modern man, ever more powerful and ever more dissatisfied. For years he pursues the project. He becomes wealthy and successful. In the final scene, he stumbles out on stage, now elderly and blind. Hearing the sound of his men digging, he feels happy to think they are 'reconciling land and water':

> To drain that pestilential swamp would be
> A final, crowning victory.
> It would open up to millions space

In which to work, possess a dwelling place;
If not entirely safe, yet safe enough
For men to lead a free and active life.
Fields green and plentiful! Men and beasts at ease
Upon this newest earth, a second paradise.

A prosperous future liberated from a barbarous past, from dumb nature. It is Peter's city, Frederick's reclaimed province, the Dutch polders. But there's a catch. Years earlier Faust had made a pact with the devilish Mephisto: the moral capital for all his endeavours had been raised from the dark side. He always knew the debt would be called in. As Faust is speaking, Mephisto explains, in a stage whisper:

These dikes and dams, they're only something good for
The demon Neptune to gulp down with pleasure.
No matter what you do, it's hopeless

Sightless Faust remains in a state of ecstatic hope: 'To see such life, such glad activity! / To stand with free men on ground that is free!' But the industrious clink of spades he could hear was not the sound of a dike gang. It was the digging of his grave.

'The soul of Faust – is the soul of Western Europe,' claims the Russian critic Nikolai Berdyaev. 'This soul was full of stormy, endless strivings.' Europe is defined by its striving – striving for truth and purpose, for control of nature. At the time of the Enlightenment and the Scientific Revolution, the continent took a dramatic leap forward. 'What ended the endless strivings of the Faustian soul?' asks Berdyaev. 'To what did they lead? The Faustian soul led to the draining of swamps, to the engineering art, to a material arranging of the earth and to a material mastery over the world.'

Berdyaev saw the finale of Goethe's *Faust* as a parable for the inevitable decline of Europe, the doomed materialist path it had set out on, spiritual energy draining away with the waters of the swamp. But it stands too as an eco-parable. Faust's conversion of the swamp had noble intentions, providing a home and prosperity for his citizens, just as the Dutch polders did. But Mephisto always meant to exact a price. Peat bogs are now understood to be one of the earth's great carbon-caches, holding in total 44% of all soil carbon. Drain them, burn them, extract the peat and calamity will follow.

Here is Mephisto:

> Your schemes, your energy, all useless;
> The elements work with us in close union
> Whose end result is death, destruction, ruin.

The Biesbosch is an epilogue to this Faustian tale. Control has been relinquished, the land left to its own ways. It is a place for scientists to study systems and species, to examine their water samples and swampy pickings on the galvanised surfaces of the Biokeet, to understand the processes of nature rather than manipulate them.

As Faust dies, he utters the famous line, which contains in it the core of our mortal tragedy. He asks for the happy moment to stay: '*Verweile doch, du bist so schön!*' (Linger awhile, you are so beautiful!).

You cannot hold onto joy – you cannot hold time, and you cannot hold back water.

I rose at dawn and stood outside the Biokeet. The sky was pale, the air full of birdsong. Blackbirds, thrushes and a nightingale, all asserting themselves with their showy syncopations and melody; behind them the passerines, shriller and fainter but audible to their own kind. Then there were the fowl, less musical – parping geese and duck, and somewhere in the hubbub, behind the party-gabble that filled that sylvan space, was another sound, deep and strange: not like a bird at all but rather something going on inside antiquated pipework – the booming of a bittern.

Later I met Rob, the park ranger. 'I don't like to pat myself on the back,' he said, reaching over his shoulder to pat himself on the back. 'But it was my work clearing the ground that brought in the bittern.'

Rob and his family lived in a proper house, one of the few in Biesbosch National Park. But it was only accessible by water, and each morning he and his wife took their children upriver by boat to a small dock where he kept a car for the school run.

I told him I was going out to meet Wim van Wijk at Ottersluis.

'Not in that.' He was looking at my canoe.

'Why not?'

'The barges. The big barges are non-stop. It's like trying to cross the autobahn on your feet. Take my bike – it's in the lock-up. Cross the Nieuwe Merwede on the car ferry.'

So I paddled to his lock-up and dragged the canoe up the bank, and set off through the polders on his bike. Across the river, in the national park, everything was wet and wild, tufted and unkempt, a sacrificial floodland now chattery with birdsong. On this side was the opposite – heavy tractors, managed fields, songless skies. It was the final vision of Faust, imprinting itself behind his sightless eyes: control, drained soil and high dikes enclosing straight-line fields and straight-line roads. Machinery tended areas of low crops, which were all the same height, all the same green. For how much longer?

I spotted a lone figure far off down the road. He was growing as he came closer. It was a rollerblader, swaying his upper body, kicking out with each stroke in a lovely liquid motion. I watched him pass, hands clasped behind his back, shrinking again.

On the skyline, I could see a large stand of poplar ahead. While everywhere else was greening up, the trees were leafless and black. The water table was rising and saltwater was beginning to encroach on places it had not reached before.

Wim stood at his garden gate and his hand was raised in greeting. The downward arc of his moustache covered his mouth; the smile was all in his eyes. We started talking at once – about my route here and what I had seen, about De Dood and the state of the side-creeks, about the fluctuating water levels, about the *place*. I've always loved those places whose physical presence dominates every moment, where the weather matters, where overt constraints and little wonders fall on everyone and begin each conversation and give common cause to all who happen to be there. Wim and I both had our own enthusiasms and observations to share – me with a whole day and a half of experience, Wim with a lifetime.

More than a lifetime in fact. His family had been tied into this half-flooded world for generations. One of his great-grandfathers had helped run the salmon fishery, overseeing a team of men who handled two-hundred-metre nets from a steam towboat. Wim recounted how, in the late summer of 1913, they took the towboat downstream for a function.

'It was a big day of celebration, the centenary of driving out Napoleon! They spent most of the day there. On the way back the towboat was in some sort of collision. Four men drowned – one was him, my great-grandfather.'

His son took over and ran the fishery. Wim's father would recall those days, his boyhood, full of boats and fishing and messing around on the banks with the fishermen. 'One day,' said Wim, 'they gave my father a ride. They placed him on a giant sturgeon. He rode around on its back. I don't think it was a very long ride, but he didn't forget it.'

The fishery closed in 1932 and Wim's parents were married after the war. Then, in February 1953 came the *Watersnoodramp*. The North Sea surged and breached the flood defences on each side – East Anglia and the Netherlands. Over two thousand people died. Wim's parents were living in a cabin by the river, and suddenly there was no river. Everything was water. Wim's father was swept away in an old rowing boat while his mother was rescued. She was pregnant with Wim and thought: *He will grow up without a father.* Hours later, her husband appeared, his hands raw from rowing.

Wim lived in a low wood-clad building, an old barracks built originally for those on the salmon fishery. Helen, his wife, joined us as we talked. She spoke Dutch with a Dutch accent and English with a Yorkshire accent. Her mother was from Bradford and had met her father when he went over to the UK to teach art. Helen had worked in the Dordrecht library as an archivist, and soon the table was filled with maps and books. Between them, she and Wim answered my questions about the centuries of water management; the yo-yoing of levels in the Biesbosch; the villages, the graves and prosperity all buried under the silt; the pre-eminence of Dordrecht as a medieval European port, and what it might have been like now but for the event that divided time here into before and after – the catastrophe of the *Elizabethsvloed* in 1421.

'We have very little from before the floods,' explained Helen. 'Only later maps and accounts that imagine how it was.' She opened an historical atlas – compiled by Wim – and pointed to the reproduction of a map from 1537. It showed what was remembered of the watercourses pre-flood, an attempt to work out what belonged to whom. 'The original map is a roll four metres long.'

There was something Old Testament in the folk memory of the flood. The destruction and its aftermath were sharp in the records but beyond them was only the dream-like awareness of a time before and the sins that caused God to wreak destruction. The sins that were most conspicuous were *moernering*, 'swamp-gaining', and *darink-delven*, 'peat-digging'.

'Everyone could see how dangerous the digging was,' said Wim. 'The weakening of the dikes was obvious. In 1385, the city sent an expedition out to stop it. They went to Zevenbergen – the name is a joke of course; "seven mountains" – they were just very low hills!' Wim's eyes lit up above his moustache. 'It probably worked for a while but in 1404, there is a record of Albrecht van Beieren again trying to stop the *darink-delven*. Then the dikes started to collapse and soon after came the *Elizabethsvloed*. The peat and salt were just too valuable.' He raised his arms in a gesture of despair. 'It's the same now, the way we use fossil fuels. We know it's bad but we just keep on doing it.' There was silence for a moment. 'We try, but is it all too late?'

I looked out of the window, out over the canal where his own aluminium punt was tied up to a post with its electric outboard on the stern. Beyond that, a few shiny-flanked horses tail-swished and grazed in the polder, and then a line of pylons and beyond them the flues and steel-clad sides of Dordrecht's waste-to-energy plant.

At dusk we went out to look for beavers. Cutting inland, we entered a rough area of reeds, slightly below the main canal. Beavers had been re-introduced some years ago from the old East Germany. Wim, as a journalist, remembered the journey to collect them. 'We had to go into Communist Europe to find beavers.'

We stood on the bank as the light faded, waiting. In the half-darkness came a heavy splash and a patch of fur slid through the water. And another. We watched them until it was fully dark. 'This is theirs now,' mused Helen. 'It belongs to them.'

We walked back, single file along the riverbank. Smoky clouds shadowed the wide waters of the Nieuwe Merwede. The traffic of shipping had slowed with the coming night but still it came, a stream of high-stacked ships heavy with the stuff of contemporary life. I could see the pinpoint stars of their navigation lights far off, and within moments they had passed us.

Wim and Helen's guest hut was a more comfortable version of the Biokeet. One whole wall, facing the river, was glass. When I woke in the morning, I thought I was in a multistorey car park. Brand-new Opels, Volkswagens and Audis filled the window. Then came the wheelhouse and coach-roof of the barge, and the propeller-bloom of white water at its stern, driving the great engine of European trade.

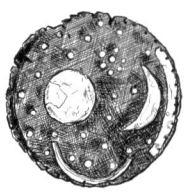

4

BRONZE

TAKE NINE PARTS COPPER AND apply heat. At just over 1,000 °C, the copper will glow and turn to liquid. Add one part tin and watch it liquefy. Pour into sand or clay moulds. Allow to cool. Brush to remove carbon. Polish. The result: a deeply gleaming metal, suggestive of gold but without gold's luxurious yellow, without its give. The tin–copper alloy is firm. It has work to do.

The discovery of bronze changed everything. From South East Asia to the peat-rich fringes of Ireland, Sweden to sub-Saharan Africa, the Aegean islands to Mesopotamia, South America to China, the miraculous alloy was everywhere. This was the beginning of the second millennium BCE, and in Egypt, in the Alps, in the palace courts of Crete, those with influence wanted to get their hands on it. They found different jobs for it, and each job was made easier, made quicker – and with each successful task they saw their status rise. The metal transformed relationships – with their rivals and allies, with the land around them – and they wanted more of it.

Copper had been in circulation for centuries, but bronze was a leap forward. Everything copper could do, bronze could do better. Bronze was easy to melt and recast. Any bronze object had the potential to become something else. A shield could be a sword, a spear, a torc. Bronze gave to its bearer speed and strength and power, as a horse did to its rider. Weapons of war became heavier and sharper, more deadly, more prestigious. Adornments grew in complexity and beauty. Farmers ploughed faster, trees were felled and logged quicker, bronze mattocks dug mile upon mile of irrigation channel. The landscape bore the mark of change.

When oxidised, bronze loses its brown-gold hue and grows a green patina. But it does not degrade. Iron – which is harder to smelt and whose widespread use comes hundreds of years later – rusts and crumbles in decades. Bronze survives. Every year, more bronze hoards are discovered and dug up, and analysis of them grows technically more and more efficient, probing deeper into questions of provenance and use. Out of the evidence – what was placed where, what was buried with whom, the ore sources and stable isotopes – appear glimpses of life, of humans acting and interacting, asserting themselves over others, over the world around them.

Bronze expanded the reach of leaders and rewarded the ambitious. It allowed surpluses of food. Disparate peoples coalesced into larger groups, into city states and civilisations. Sumerians, Babylonians, ancient Egyptians, Minoans and Mycenaeans, Hittites and Hurrians all had bronze at the core of their lives, as did the Oxus culture further east and the Xia and Shang dynasties in China. When the great Shang general Fu Hao died, 1.6 tonnes of bronze objects were buried with her.

Now three figures step forward. They come to the front of the stage with their three new skills for the new age: smith, warrior, trader.

Sweaty from his cabin of secrets, hammer in hand, bare-armed and leather-aproned, the metalworker is high priest of the new technology. He alone is privy to the rocks' powers, can mix and heat them in the right way. All that can be gleaned of his work in the forge is smoke and the sound of hammering, ceaseless hours of hammering – not like the hammering of stone but something sonorous and musical. The Buryat and other Siberian groups believed the smith's faculties stretched beyond his metalwork and were cosmic and transcendent, trumping even those of the shaman if it came to a contest of souls. In ancient Java, the prince and the smith were seen as brothers. Native American metalworkers passed on details of their craft only to close family members. Wherever bronze was made, guilds of smiths grew up, closed societies, keepers of mysteries.

Sometimes, like the Asur of India, smiths were outcasts and sometimes they were revered. They could recite genealogies and their art was likened to that of the epic poet, the musician. The smith could produce something akin to the spark of life and his furnace was likened to the womb. In it, he re-enacted the divine act of creation.

To the Greeks, he was a god – Hephaestus, son of Hera and possibly Zeus. Squat and limping, Hephaestus bore the scars of his trade. In the absence of tin, arsenic was added to strengthen copper, and those who worked with it were affected. Norse, Ugarit and ancient Egyptian mythologies all feature a smith god with similar impediments. Hephaestus fashioned the weapons and thrones of Olympus. It was from Hephaestus' forge that Prometheus sneaked the gift of fire – and it was Hephaestus who was ordered to beat the shackles closed when Prometheus was punished. From the seed of Hephaestus and the womb of Gaia came the great city of Athens.

In *The Iliad*, Homer tells us how Hephaestus is asked to make a shield for Achilles. He commands his twenty bellows to start blowing. He fires the metal, places it on the anvil, and what he produces is something more than a shield, as metal is more than stone. Beneath his hammer the bronze flows and flexes, dynamic and fluid in its gleaming complexity, and he recreates a series of scenes so elaborate that no physical object could actually contain them. There is something crazy in Homer's description of the work.

Hephaestus shapes a bronze cosmology: 'earth, sea and sky, the tireless sun, and the moon waxing into her fullness, and on it the constellations'. He adds in the stars of Orion, the Bear and the Pleiades. He shapes the images of two towns. In one a dispute is underway, with cheering supporters and a ring of elders who sit around a pile of gold, the prize for whoever resolves the dispute. And in that town is justice and advocacy while in the other town is war, besieged as it is by soldiers in glittering armour. The towns-folk stage an ambush but the enemy counters with chariots and bronze spears, and there follows a terrible battle in which Strife and Panic and the dreadful Demon of Death mingle and clash. All this in bronze. Next, Hephaestus makes a scene of ploughing – 'a miraculous piece of work' – and a panel of reaping, of sickles and sheaves, and a feast, and an image of a grape harvest, and another of a magnificent herd of cattle, and another of sheep-flocks grazing in wonderful pastures, and a dance floor filled with wheeling groups of girls and boys and solo dancers – and around it all, around all that movement and life with its glories, its plenty and its horrors, he crafts the mighty Stream of Ocean. The shield is the material world made animate, blooming into life through the medium of metal. Only bronze could carry such scenes, only the hand of the smith could forge them.

Beside the smith is the second figure: his client, the warrior. Bow-chested in bronze cuirass, crowned with a horned bronze helmet, spear-flanked and greave-legged, he stands splendid. Violence itself is nothing new. The warrior's stone- and flint-weaponed ancestors had the means to kill and did so. But bronze now brings culture to conflict, giving rise to a dedicated class of fighters and embellishing them with the finest products of the forge. Valour on the battlefield becomes the highest of virtues. Deeds of glory and courage live on in stories, told and re-told, shaped and polished into epic. The hero is born, the very model of mortal perfection. The bloody reality of fighting – hacked limbs, axe-cleaved flesh, torsos punctured by bronze spears – is subsumed beneath the ideal of the conquering male. The warrior is beautiful as well as brutal, the burnished man in his burnished armour, terrifying, impossible not to gaze at. He maintains his appearance as well as sharpening his weapons. Alongside bronze swords and bronze axes in warrior burials have been found tattooing awls and mirrors, tweezers and razors and combs.

Bronze was hard to source and hard to forge. It produced objects that were highly prized. So it's a surprise to discover that many bronze items were simply thrown away, unused. Hoards of axe-heads and swords were buried in pits. Some hoards were vast – a group of six in Transylvania totalled ten thousand pieces of bronze, weighing five tonnes. At Isleham in Cambridgeshire, a six-thousand-piece hoard was unearthed and many of the pieces still have flashes and seam-spillage on them. They were purposely disposed of, often in significant places in the landscape – near water, for instance, on hilltops, or at liminal sites beyond the boundaries of settlements. A clue as to what lay behind this practice can be found at Stonehenge, where some of the sarsen stones had little bronze axes painted on them, down at ground level, as if they were rising from the soil. Metal objects returned to where the metal came from, to nourish the forces below and ensure future supply, like seed-corn taken from the harvest. Perhaps doubt was also a part of it, the sense that bronze is not deserved. It belongs in the earth.

Beside the warrior and the smith stands the trader. Mule-driver, sea-captain, middleman or prince, he takes many forms. In Ugaritic texts of the time, there is talk of a caste of wealthy merchants known as *tamkaru*. Cuneiform sources speak of caravans with hundreds of donkeys trekking between Mesopotamia and Anatolia. In the quest for tin, ships pushed further and further along the coasts and up into the sheltered rivers of Europe – the Fal, the Humber, the Gironde, the Rhine. Others sailed out from the Mediterranean and up along the Atlantic shoreline, or down the Red Sea and across the Indian Ocean. There had been movement before the Bronze Age, of people and flints, stones and beliefs. Now it was ramped up. Raw tin and copper and finished objects of bronze were carried in bulk for thousands of miles; there are accounts from the period of mule trains three hundred strong. Never before had distant human groups interacted on this scale.

From the Biesbosch, I paddled back to the Vissershang marina and from there travelled by various trains to the Dutch river port of Venlo. It was Sunday morning when I reached the gates of the Geurts-Janssen steel depot. There was no-one about. Rain skittered across the empty concourse and rang out from the corrugated roof of the warehouse.

I spoke into my phone. 'I am here, Franz. Outside.'

Franz was on the inside, beyond the warehouse, waiting at the wharf on the bridge of his eighty-metre barge, the *SS Seestern*. 'One minute! I am coming.'

Franz and I had had a lively correspondence about where and when to meet. His texts alternated glimpses of life aboard the *Seestern* with a constantly shifting itinerary. Come to Homberg, he wrote. No, not Homberg – I have a storm here and I'm stuck in Gendt. Now the cover of my lifeboat – it is ripped off! I must go up the Moselle. I am struggling in the wind. War in Ukraine has sent schedules into chaos . . .

'I get an order,' he wrote. 'Five minutes later it is cancelled. Shipping, it's like a running bet on horses. You cannot predict.'

And now a figure was approaching, hooded and hunched against the rain. At the gates, he tapped away at his phone, struggling with the code. The barriers lay between us like an unmanned border. I could hear him muttering as he tapped. Then, with a shudder, they began to rise.

'*Sesam, öffne dich!*' Franz stretched out his hand beneath them. 'Open sesame!'

He led me through the main building, a cavernous emporium of metal. Steel-sided and steel-roofed, its unlit galleries stretched out into darkness. The floor was covered with stacks of steel girders and steel mesh. Lying on the wharf outside, below the gantries, were more rows of building steel, rods and sheets, tinted rust-orange, shiny in the wet. Against the quay beyond them, with a hold full of steel, lay the steel length of the *Seestern*.

Over the coming days – with Franz, his wife Sonja, Stella the ship's dog and a cargo of eight hundred tonnes of finished steel – I pushed through the waterways of the Low Countries, up the Rhine and into the heart of the continent. If modern Europe is a living body, then these rivers and canals are part of its cardiovascular system, the veins and arteries carrying what is needed for its elaborate life – raw materials and fuels, coal and natural gas and iron ore, cars and consumables. It is a highly evolved and complex system but in essence no more than a scaled-up version of Bronze Age trade.

I'd long had a fascination with these riverboats. I'd watched them from the banks of the Rhine and the Danube, sliding under bridges in Cologne or labouring against the current near Vienna or in Boppard. They bring out the same curiosity and yearning as ocean-going ships – tugging at the soul with the compulsion to clamber aboard and head for the horizon. But river barges do their freighting close at hand, their long low shapes just a stone's throw from shore. After you've watched the hull pass, and the wheelhouse and after-cabins come into view, you see potted plants in the windows, and on the coach-roof above, a link to the land – a car. The *Seestern* was at the smaller end of the scale; on its stern was no car, not even a bicycle, nothing but Franz's Australian barbecue.

'Ach, the rain!' Franz trotted up the steps, kicked off his crocs, and entered the wheelhouse. He took his place in the swivel-chair, before an array of screens and dials. 'We wait.'

74

He had a warm, open face, pepper-and-salt hair. A small gold anchor dangled from his left ear. Six generations of bargemen lay behind him. 'Perhaps it is seven. I don't know exactly.' He was German and his first language was German. 'But the rebel blood of the French is in my veins!' he said, face suddenly aflame – and I believed him. 'In 1851, my great-great-something grandfather was Minister of Defence in Paris.'

Franz's own life had been all boats. He had been born and brought up on his parents' barge – also called the *Seestern* ('starfish'). Being itinerant, they sent him to boarding school at a convent in Mannheim. He shook his head at the memory. 'It was not a happy time for me. Some of the other boys were quite rough. I liked to play the piano.'

For a few years in his twenties, Franz worked on shore, in a bank. 'Each day I went in and I looked at numbers. It was more like a holiday than work – not like this.' He gestured at the control panel in front of him, the long hold below the window, the bow beyond. It was clear which he preferred. Since then, his land-life had been more or less extinguished. 'I have no passport – no room on shore, no roof except for this. This is our home and our office.'

Sonja came in with coffee. They had not been married long; each had grown-up children by other marriages. Sonja was not from a boat family. I asked her how she found it.

'I like it.'

Franz raised an eyebrow.

'No, really!' she laughed, handing me a mug. Together they were able to handle the boat in the trickier manoeuvrings and moorings. 'You'd be lost without me.'

Franz gave her a teasing smile. 'It *is* useful having another pair of hands on deck.'

The rain intensified. We could hear it tapping overhead, see it puddling on the companionway outside the door. I watched it bulge into tiny bun-shapes on the handrails. The water of the dock was grey and spotted with drop-rings. Everything else was Sunday-still – cranes and silos and sand-piles.

I became aware of a splashing sound. From onshore a high gutter pipe was spewing water onto the concrete, the steel piles and the side of the ship.

Franz was talking about the brass bell which hung just outside the wheelhouse. It was a legal requirement to have a bell on board, with a statutory metal thickness and size, as an alarm of last resort. Franz's bell was well polished and engraved with SEESTERN 1928 – it came from his father's ship, the *Seestern* of his childhood.

'Every time we leave port, I ring the bell. I make one ring, and a pause, then two rings, and a pause. My father always did it. *Gute Reise* from God!' I couldn't tell if he was serious. 'A few months ago, I forgot. Did it matter? The ship went backwards into the side! Look on the stern and you will see a very big dent. I will show you when— *Scheiße!*'

'What?'

'Electrics!' He was already halfway down the ladder. I followed him into the engine room. Water from the high gutter was pouring in. Franz peered behind the bulkhead where he kept his fuse boards; there were little streams running down inside the hull. '*Scheiße – Scheiße!*' But it was OK – somehow they all missed the wiring. Up on deck we placed rolled-up cloths to steer the water into the scuppers.

The rain cleared. Franz gave me a pair of work-gloves and I went up into the bows with Sonja and we each took one of the heavy plaited warps. She had a headset on and was speaking with Franz far astern. The bell rang once, then twice.

'OK – *jetzt!*' urged Sonja.

We flicked the lines off the bollards and hauled them in. The gap between ship and shore widened. The *Seestern*, with its burden of steel, crept out of the docks and into the river.

All morning we headed down the Meuse. Low banks and grazing cattle, weekend anglers static beneath umbrellas. By mid-afternoon we'd reached the Rhine. Under clear skies we turned right and began to push into the flow, against the weight of water, uphill.

I found an old deckchair up in the bow, just in front of the capstan. The anchor tubes gurgled below me; the stem's unseen cleaving made a constant water-swish ahead. I sat with my boots up on the steel coaming and felt happy with the thought of the days ahead, locking into the slow boat pace, contemplating the great river. I had been studying German and could now practise with Franz and Sonja. I had a stack of books.

The river was running fast, heavy with the distant streams of Alpine snowmelt. It didn't flow smoothly but in a surface of brownish eddies and upwellings, like a million dissenters all jostling and hurrying – but all marching in the same direction.

Barge after barge pushed past. Some had black mounds of coal running along their backs. Others looked empty but hid denser dumps of iron ore in their hulls. There were tanker barges – sealed holds and a tangle of pipes on deck. One had a floral design on its bow and in English, a slogan along its topsides: KEEP OUR RIVERS CLEAN! (Do *you*? I thought). Several were piled with scrap, the waste metal of our age, tangles of crushed and broken steel, off to feed the cannibal blast furnaces, to produce new steel for new projects. A double-decker overtook us – with two storeys of brand-new John Deere tractors. Stacked together, they looked diminished, toy-like in their livery of green with yellow wheel-rims.

*

Europe's most celebrated piece of ancient bronze is about the size and shape of a vinyl LP. Its colour is a lovely green patina. While one side is blank, the other is inlaid with shapes of gold to create an image of the night sky. The Nebra Sky Disc distils three of the main concerns of the Bronze Age – the distant mysteries of the cosmos, the physical mysteries of metal, and long-distance trade. It is the earliest known representation of the sky anywhere, and its fascination lies in part in how a small piece of bronze can embody so much.

The disc displays a gold moon in two phases, full and crescent. Gold stars are scattered over the surface, including the seven Pleiades. When the disc was first smelted and forged and decorated in the early Bronze Age, the Pleiades would have disappeared from the night sky about 17 October, with a new moon, and then returned on 10 March, with a full moon. This second moment signalled the start of the agricultural season. The disc appears to address one of the great questions of the time: how to sync the passing of days with the change of seasons, how to know when to plant crops? The Pleiades and the moon could also be used to calculate leap years.

Running along opposite sides of the disc's rim were, originally, two gold bands. These strips were added later and they each occupied 82° of the circle – exactly the proportion of the east and west horizons over which the sun rises and sets throughout the year. Such precision suggests a functioning sky-guide, a version of the classic Philip's Planisphere with which every child-astronomer was once familiar. But what the Nebra Sky Disc shows is not complicated. It would have been understood by any group who kept their eyes on the night sky – and there was not much else to do after dark: the day was sun and activity, the night moon and signs. The disc might have started as a tool, but over time probably became less functional than symbolic, the embodiment of a cosmic law, a physical object made sacred by its manifest truth.

Towards the bottom of the disc, there's another golden arc. It looks somewhat like the smile on a smiley face, but it is thought to represent the 'solar boat' which ferries the sun daily across the sky. Notches along the arc's side are oars, or possibly rowers. In Denmark, a hundred miniature gold boats were discovered in a clay vessel at a Bronze Age burial, their tiny hulls inscribed with sun symbols; countless rock paintings in Sweden show sun-boats, while a number of decorated bronze razors reveal variations on the same theme. The gold sun-boat was added late to the disc, and its meticulous feathering and fine craft embellished an already revered image. The whole thing has a beauty without scale. It's a portable monument. The more you look at it, the more extraordinary it appears.

Like many Bronze Age artefacts, the end of the Nebra Sky Disc's career came not with obsolescence but with a ceremonial burial. At some point around 1600 BCE it was deliberately removed from circulation, its cosmic images confined to memory, placed back in the pocket of the earth. It was taken to the top of the sacred hill of Mittelberg and slotted upright between stones, buried standing. Other bronze objects were buried with it – an axe-head, two swords and a cluster of armlets, laid before it like flowers at a headstone. Then it was covered in soil.

We do not know how many people saw the Nebra Sky Disc, but from DNA analysis we do know more generally how far some individuals travelled during the Bronze Age. It was a period of movement and connection. In her 2015 essay, the Danish archaeologist Helle Vandkilde came up with the term 'bronzisation', globalisation in embryonic form, and the idea has caught on: 'the Bronze Age . . . constituted a unique case of pre-modern interconnectivity . . .

an overarching globalising phenomenon, tightly knit by one crucial resource'.

Bronze opened up the world – its range, its scale, the sheer richness of its resources. The rarity of tin and the lengths required to obtain it acted as a stimulus. Evidence from the period suggests a growing appetite for combining and transforming the materials of the earth, and a more general spirit of novelty and innovation. To the Bronze Age we owe the introduction of writing, the wheel, legal codes, civil administration.

Trade generated trade. Along the same routes as ores, metals and inventions flowed exotica. 'Bronze Age mobility seems firstly motivated by the increasing dependence on metal,' wrote Vandkilde, 'and secondly (as had always been the case) desire for rare materials.' Colour and flavour followed ingots across steppe and desert, sea and river – mainly from east to west. Blue lapis reached Babylon and the Levant from Afghanistan, honey-orange amber was carried south from the Baltic, cherry-red carnelian from India. Melons, citrons and cucurbits from South East Asia reached Anatolia, likewise cloves and nutmeg and jasmine. Peppercorns from India were used in the mummification of Ramses II. So intrigued was the Egyptian Queen Hatshepsut by traders' offerings that in 1493 BCE, she herself travelled down the Red Sea to the Land of Punt, returning with gold and rare woods and live apes and trees uprooted to be replanted in home soil. Aegean frescoes of the eighteenth century BCE show grey langur monkeys from India. The chicken (*Gallus gallus domesticus*) originated in China and first appeared in the eastern Mediterranean at this time.

Trade was not simply the exchange of goods. The journey itself was a commodity. Mary Helms is an anthropologist who has studied the importance of geographical space in pre-modern society. She suggests that those travelling for trade returned with knowledge and experience that gave them 'an aura of prestige and awe'. Bronze itself was accorded something similar, a layer of significance added by its diverse and far-off origins. 'Bronze artifacts,' she writes, 'were further empowered by the qualitative "difference" that may have been accorded by society to things that came from places cosmographically "out there", beyond the horizon.' Looking at the practice of burying precious bronze, Helms remarks that 'the sheer human effort required to obtain metal from the greater distance further enhanced the overall supernatural potency of the deposited bronze'. The labour of travel added weight to bronze.

The Nebra Sky Disc holds layers and layers of meaning and one of those layers is distance. In its materials, in its bronze and gold, is earthly distance and in its image the limitless expanse of the sky.

Franz at the helm, one socked foot on the edge of the instrument panel before him, leaning back in his swivel-chair, left hand fiddling the electronic helm, an eye on the foreground screens (chart-plotter image, the river, our position, all the other named vessels nearby; four-way split screen of CCTV cameras on the bow and stern, port and starboard, separate screen of the engine room).

There was no autopilot on his river barge. In the open sea, you can key in a course and the autopilot will keep the ship on it. But the river offers narrow waters, no straight lines and a constant parade of oncoming ships. The maritime convention of passing portside to portside is followed, except that from a certain point upstream of Bonn, it isn't. We moved into the Upper Rhine, where traffic has the right to seek out the slower streams of the inner bend. If they want a starboard passing, these struggling barges drop a blue board beside the wheelhouse and then the approaching ship knows. All the machines and electronics are no match for eyes and judgement.

'My problem is this,' explained Franz. 'I am tired. I do not sleep well at night. I have tablets but I hate to take them. So I'm tired.'

He could not leave the helm. From the darkness of early morning to sundown, he was there. We had breakfast there, in the wheelhouse, and lunch. Sonja could relieve him for short periods, and I took my hand too, but Franz still had to be around to operate the short-wave radio, to make all the big calls. At night we tied up alongside wharves and small harbours and we'd eat in the galley and then retire. Franz was spending his down-time watching a boxed set of the BBC series *The Onedin Line*. Sometimes there would be fuel to take on in the morning, or loading and unloading. At other times we left with the sun rising, or in the dark.

One night, with the lines and springs looping from the ship to a harbour wall, Franz said he had watched the last episode of *The Onedin Line*. He poured schnapps into tall, twist-stemmed glasses and produced a photo album. Its burgundy cover was worn and age-faded, like a family Bible. He turned the pages, lifting the tissue interleaf, and each spread was a series of tiny monochrome images, whole lives caught within the borders. Here was his uncle, who resisted the Nazis and was sent to the Eastern Front and never came back. Another uncle died in the war. There were pictures of the first *Seestern*, in a lock, on the river with a cargo of scrap metal, being repaired. Strange how similar the designs of the two ships were.

The sepia images were thick with nostalgia. A couple arm in arm, in formal garb; a family group in the ship's bow – three generations with the sun in their hair, legs dangling over the side. Then a harbour with a mass of tugs – pre-war. A great deal of cargo was carried in vessels without their own engine, and the original *Seestern* began that way too. Passage on the waterways relied on these small craft. In the photograph, there were dozens. Their sheer number was a spectacle, a glimpse of the extent of river commerce. But they were in harbour, not on the river working. They were redundant. It was the early 1930s: the German economy had taken a nosedive, and the global economy with it.

In another picture, a flare of white light is saturating the image, but I could just make out the ship's wheel and a boy's face caught in a doorway.

'Yes – it is me.'

And here again, a stout man alone on deck. He wore a pale jacket, too tight on his rounded frame, as if his bulk had come on suddenly, straining at the buttons. There was something rock-like in his stance, and it was clear that here at least – on the foredeck of the *Seestern* – he was master.

'My grandfather,' explained Franz. 'They called him *Galo*, "cockerel".' He smiled, making a flapping motion with his elbows. 'He must have had a hundred wives.'

We delivered the steel in the morning and pushed on upriver. The next order was sand for a building project up the Moselle. Just below Duisburg, we turned into a small gravel-pool beside the river, a *Kiesloch*, with a large gravel dredge moored to the inner bank. We tied up alongside it, ready for loading in the morning.

I went with Franz down to flush out the hold. The high sides boxed in the sky above; down here it was the size of a couple of tennis courts. Our steps echoed metallically. Towards the stern, the last of the ballast water was being pumped out – and we stood watching it.

Franz scooped up something from the dregs. He opened his hand and in it was a tiny perch, no more than five centimetres long. I could see now there were others – pale flecks on the hold floor, unmoving. I picked one up and held it in my palm – its sightless eye

83

gazed up from the hold, its moon-coloured scales still shimmered. Each little fish had been sucked up through the pump.

At dawn, the hull of the *Seestern* gave a shiver. Beside us the dredge whirred into action. A dozen conveyer belts began to run, while below the water a row of belt-buckets scraped at the river-bed. The material worked its way up, passing through sifters, which shook out the coarser gravel and channelled it onto different belts to load a barge on the opposite side.

We were in the wheelhouse. Off to the east the sun was rising behind the great shoulders of Hermann Wenzel power station. The underside of the clouds glowed with a pinky-red that looked artificial. A crowd of black-headed gulls flitted astern, their shapes milky in the half-light. All the while, the hold was filling with sand the hue and texture of demerara sugar, dropping from the swing-arm to an appetising cone, whose slopes rose and rose until they reached their tipping point and collapsed. On either side of the wheelhouse was a level-light, which flickered and flashed if the load was not centred and the ship began to list. Then Franz would radio ashore to adjust the arm.

Of all that we take from the earth – the ores and oils, copper and clay, gems and coal, bauxite and black shale – only water exceeds sand and gravel in volume. Fifty billion tonnes annually, destined to become concrete. With it, you could build a wall around the equator twenty-seven metres high and twenty-seven metres wide. It's a geological process on a global scale: vast water-filled scrape-holes dug out in places like this, cities and bridges and tower blocks forming as features elsewhere. In time, those structures will collapse and pebbles of concrete will be washed to the sea, rolled as rounded clitter on the riverbed, ground back to constituent grains, pressed again into sediments.

Exiting the *Kiesloch* was not easy. We had to head out of the entrance and briefly turn broadside to the stream. The river needed to be empty of ships and the manoeuvre had to be done at speed lest the full flow of the Rhine catch the bow. Franz was nervous. His head was constantly in motion, checking downstream for traffic, checking upstream, checking the plotter. He pushed the throttle and twisted the helm hard. The long thrust of the hull swung right, moving fast against the backdrop of the far bank, then slowed. The river was clear; we had the room to straighten. We were back in our slot, now laden and low down in the water, the stern of a barge far ahead and two more coming down. I gave Franz an admiring thumbs-up.

He eased off the throttle, pushed up his glasses and rubbed his eyes. 'What's the time? Nine?'

'Half eight.'

'I'm already tired!'

Whatever mystique long-distance trade once had has now gone. The idea of bronzisation overlooks one difference: rarity. What was carried along trade routes then was unusual. Now, at every moment of every day the waterways and roads, the railways and air space, are full of bulk commodities in transit. Few of them surprise us. The *Seestern* is a tiny cog in that machine. The margins Franz and Sonja work for are getting slimmer, costs rising, yields falling. The generations of river families, the long lines of barge-captains, have come to an end. Franz's daughter, his only child, is the commercial manager for a biscuit company.

At Koblenz, the *Seestern* took a right, into the Moselle, and that's where I left. When the hull had risen in the first lock, I threw my bag on top of the wharf and climbed up. The upper gate opened and I watched the *Seestern* enter the smaller river. Sonja, headset on, hands in gloves, waving. We exchanged a '*Gute Reise!*' Then came the hold and the sand and the wheelhouse and Franz grinning a farewell before returning his gaze to the water ahead.

Farewell, Franz! Farewell, Sonja, the *Seestern* . . . may you all get some rest.

I took a train eastwards. The next day I reached the town of Halle where, in the *Landesmuseum für Vorgeschichte*, the State Museum of Prehistory – after an eventful journey of its own – the Nebra Sky Disc has ended up.

A hot afternoon in July 1999. Henry Westphal, a road worker, was busy with some weekend metal-detecting on the Mittelberg hill, near the small town of Nebra in the east of Germany. Sudden pinging sent him reaching for his pick. Almost at once the blade struck – metal on metal. Digging out the muddy green disc, he realised this was something more than the coins and militaria he'd been after. He threw it on the back seat of his Trabant and went off for a celebratory drink with his friend. Within a week or so they'd managed to sell the disc for 28,000 marks. The buyer was a man named Achim S., who bought it because he knew he could sell it for a lot more. Achim S. took it home, filled a bath with water and washing-up liquid and soaked it for three days, before attempting further cleaning with wire wool.

Unfortunately for Achim S., the state of Saxony-Anhalt has a legal entitlement to all such finds. When he put out feelers for a sale, he found wariness. One who was not wary though was Hilde-gard B., who ran a bar where collectors of antiquities gathered every month. She had not yet seen the disc but it had become the talk of her clientele. She found herself obsessed. She experienced a vision and began work on a novel about the disc. At the same time, she managed to negotiate an underground sale for Achim S., and it was sold to a buyer in Switzerland.

Harald Meller was the State Archaeologist for Saxony-Anhalt. Through his own networks he learnt of the disc and started to track it down. He was keen to assert the legal right of Halle Museum to acquire it. The Swiss owner named a price of €700,000 and agreed to meet Meller at the bar of Basel's Hilton Hotel. The owner arrived with the disc taped around his waist. Now he opened his shirt and laid it on the table. Meller had prepared some harmless chemicals and set about applying them to the patinaed bronze. At that moment, others in the bar stood and closed in – they were all undercover police. The owner was arrested. Meller and the state of Saxony-Anhalt got the disc.

So began a new chapter for the Nebra Sky Disc. Tens of thousands of people came to gawp at it in the museum. During its lengthy spells in the lab, digital microscopes scanned its every micron, converting the metal into 3D images and its surface into topography. A line of chisel-scoring was described by the chemist Dr Wunderlich, using intense magnification, as a 'cross-country ski run'. Archaeometric analysis revealed that the probable source for the bronze's copper was the Austrian Alps and for the tin Cornwall – specifically the alluvial deposits of the Carnon River, which drains the Carnmenellis pluton.

Tracing the origin of the gold proved trickier. The team of forensic archaeologists tried to match it with known sources of ancient gold. They drew a blank. An archaeo-geologist whose speciality was gold was called in. As a young man, Dr Gregor Borg had shown an uncanny ability to find gold where most thought there was none. He discovered two gold deposits in Tanzania with a value today of $2.55bn and another that had been written off by geologists. Gold, he considers to be a 'warm and calming' metal, but with a tendency to disrupt: 'I have learned that there are two things in life that make people act irrationally: gold and cars.'

Borg and his team spent six years examining the gold of the Nebra Sky Disc. Its signature was micro-traces of silver, copper and tin, and they set about finding a match. They consulted a database of four thousand gold artefacts. None came close to the gold signature of the Nebra Sky Disc. They looked at natural gold from two hundred sites. Again, nothing. Then one day Borg had a call out of the blue – from a man speaking 'excellent German with a charming English accent'.

The accent was actually Cornish, and it belonged to Courtenay Smale.

I had been to visit Courtenay months ago, when I heard he had one of Cornwall's foremost mineral collections. Well into his eighties, he lives in a neat bungalow in Newquay. One whole room was given over to his minerals. Rosettes from international mineral shows ran around the top shelves. The samples he produced for me were spectacular – the psychedelics of blue liroconite and blue spangolite and green torbernite, the 'pigs' eggs' of kaolinite. He also handed me a plainer one, a pretty cluster of everyday quartz crystals.

'I take that to shows, and people say, "What's so special about that?" And I say, "Turn it over."'

I turned it over. The crystals, said the label, were given to him by his great-uncle Dick at the age of seven.

'He was a beam-engine driver in the clay country, and they would find these rocks. Spar crystals. We went to him for tea on Sundays, and the crystals were in a corner cupboard. I used to gaze at them . . . Then one time, Uncle Dick opened the cabinet and he took them out. "Have them," he said.'

Courtenay smiled a modest smile. 'That's where it all began.' He gestured at the 'all' around him: the collection cabinets, the mineral books and displays, a lifetime of methodical study and acquisition, of burrowing deeper and deeper into the earth to recreate the joy of first holding those spar crystals.

Some years ago, Courtenay was called in to catalogue the private collection of minerals at Caerhays Castle. The castle is in the south of Cornwall and is owned by the Williams family, who did very nicely from Cornish mining concessions in the late eighteenth and nineteenth centuries.

'One day I opened an old drawer in the castle and there were thirty or so gold nuggets. They were just lying there, no labelling of any sort. I noticed that one had a tiny grain of cassiterite embedded in it. Of course, I'd heard of Gregor Borg's search for the Nebra Sky Disc's gold and thought these would be of interest. So I telephoned him in Halle.'

Smale had learnt German while doing National Service in Germany. His late wife was also German. 'When I told him about the gold, Gregor was so excited he said he'd come to Cornwall at once.'

Borg arrived at the castle with a handheld X-ray fluorescence analyser. The nuggets looked promising, so they went back to Halle and were given the full examination. The unusual suite of trace metals – silver and copper and tin – was what the Halle team had been looking for all along. The problem was that the Cornish nuggets themselves were unsourced. Borg and Smale pieced together scraps of historical material – hand-drawn illustrations of the nuggets in a nineteenth-century journal, circumstantial evidence (the Williamses had the mineral rights to the area). It all pointed to the same place: the Carnon River, which flows down from Carnmenellis – the granite lump that later helped fuel so much of the Industrial Revolution – and into the River Fal.

It was hot in Halle. All day the sun had been pumping its heat down into the city – by late afternoon the city was radiating it back. The crowded tram hummed and clunked over steel rails and everyone in it looked about to scream at each other, or faint. Arriving in the thick-walled *Landesmuseum für Vorgeschichte*, I found the atmosphere instantly cool and hushed.

The Nebra Sky Disc was currently away on loan, but I'd arranged to meet Regine Maraszek, the disc's chief curator. I waited for her in the museum shop. The disc was everywhere – its green-and-gold, moon-and-stars image filled numerous book covers, and you could hang posters of it in your bedroom, have it on your writing paper, your key ring, your paperweight and mouse mat, your cufflinks and your coin holder, or you could cover your mouth and nose with a gold-and-green, moon-and-stars, sky disc face mask.

'I was lucky,' said Regine. We went to sit outside in the shade. 'I joined the museum in 2002. I came with a Bronze Age speciality and it was just the time the sky disc arrived. They put me in charge.'

'How do you explain its popularity?'

She paused. 'I can only say that there's nothing remotely like it. With the onset of the Bronze Age, you have a lot of buried jewellery and swords and things. But nothing like this. We know how connected Europe was at the time. There are many Bronze Age objects to show that. The sky disc is the only one with such an image, the only one to show the sky.'

I saw the disc later, when it came to the British Museum. Initially I struggled with an *is that it?* feeling after long expectation. When he first spotted it in the soil, Henry Westphal thought he'd found a

bucket lid. I bent to peer more closely. I examined its features, familiar now from dozens of pictures. I thought absently: it's about the
size of the ammonite . . . Then something shifted and I was struck
not by its surface elements of bronze and gold but by the whole, by
its overarching purpose, revelation of the sky.

A widespread belief in antiquity saw correspondence between
the vault overhead and the earth. They mirror each other. Understand one and you understand the other. They are connected by
universal affinity. The seven main heavenly bodies had their equivalent in known metals – sun (gold), moon (silver), Saturn (lead),
Jupiter (tin), Mars (iron), Mercury (mercury), Venus (copper).
Something similar can be found in the glimpses we have of the
cosmology of ancient Egypt, Mesopotamia and Greece. This belief
was commonly held in Europe until the sixteenth century. In *The
Canon's Yeoman's Tale*, Chaucer spells it out:

> The bodies sevene eek lo! hem heer anoon:
> Sol gold is, and Luna silver we thrape,
> Mars yren, Mercurie quik-silver we clepe,
> Saturnus leed, and Jupiter is tin,
> And Venus coper, by my fader kin!

In the same way that metals make up the deep core of the earth,
so the high heavens reflect it. In the *Kalevala*, the great Finnish epic,
the sky is metal, created by Ilmarinen the smith, who hammered
and forged it with such skill that nowhere on it is any mark of anvil
or tongs. In Pharaonic Egypt it was understood that the metal sky
was pricked with holes and at night we see these holes and they're
the stars. Common to dozens of traditional cultures was the notion
that sometimes little bits break off the metal sky and drop to earth
as meteorites.

91

Behind such beliefs is one idea that embraces all variations, all the esoteric details that spin away from them. The earth and sky are a whole – two sides of a great cosmic commonality, and the movements on earth and the movements in the sky follow the same pattern, and the materials of each reflect the materials of the other.

Out at Nebra itself, the disc has been re-integrated into the landscape. The area around the find-site is now curated. On the approach to the hilltop is a new building in the shape of the disc's golden sun-boat. Sixty metres long, clad in gold panels and articulated on a flat plinth, it is a structure whose interior logic has been entirely sacrificed to its outward design. It acts as a visitor centre but looks like a giant banana. One end of it is a vast panel of glass, which faces the summit of Mittelberg, two miles away.

I walked up through the forest. Thrushes and a cuckoo sounded from the beeches. Bud-fresh leaves glowed in the windless air. Sunlight pierced the understorey. My boots crunched in last year's foliage, and the boughs and branches overhead made the light break into a thousand green and yellow pieces: the joy of being among trees, an everything-at-once joy, of place and the fleeting moment.

Near the top of the hill are two artworks. One is by Hayato Mizutani, entitled *Waldpavilion*. It's an open-topped tubular structure, which you step inside and find your gaze funnelled upwards to a ring of cloudless blue. Not far away is a theatre signboard, conceived by Halle-based artist Michael Krenz, mocking up the announcement of a star-studded show: TONIGHT – THE SKY. Laid down on the spot where Henry Westphal's metal detector first pinged is a convex mirror – a stainless-steel eyeball which holds in it an image of the sky above. Some way off stands a thirty-metre-high tower, which leans slightly, like the gnomon of a giant sundial.

What would future archaeologists make of it all? What would they understand of the people who erected a golden arc of a building on the slope below, which pointed to the hilltop and the high tilting tower? Speculations would multiply, as they tried to imagine beliefs and ideas that lay behind the structures, which must in some way be connected to these other objects discovered in the wider region – the sky disc replicas, the cultish cufflinks, paperweights and coin holders.

Many centuries ago, a thousand years after the sky disc was buried here, a circular rampart was built around the hilltop. The people who built it were not aware of the sacred object that was propped up beneath the soil, and they never discovered it, but they did know about the strange importance of the hill's position. On May 1, after the feast of Beltane, the sun would sink behind the far-off peak of Kulpenberg. Even more significant was the hill on the skyline some fifty miles to the north-east – the Brocken, the highest point of the Harz Mountains. At dawn on the summer solstice, the gold disc of the sun was seen to rise precisely from the Brocken. Together these celestial alignments had helped build layers of sanctity around the hill of Mittelberg.

The urge to climb up here in the darkness, to witness the coming of the longest day, is an impulse we can recognise. It is a need for pattern in the shapeless landscape, for a mark at the hinge point of the turning year, something predictable in a chaotic and ever-changing world. The same impulse prompted people to craft in bronze and gold an observable rule in the heavens in the hope that it would remain true, to create a sky disc in the finest materials available, to revere and cherish it with all due solemnity, for generations – then affirm its eternal value by burying it away in the ground.

5

SILVER

SECOND BEST. BETTER THAN BRONZE, but always runner-up to gold. Never the winner. Shy reflecting moon to the burning gold of the sun. In Hesiod's Silver Age, life was markedly less pleasant than in the Golden Age that went before, when the earth had provided its bounty without labour, and all mortals lived in languid contentment, freely mixing with the gods. Eras of artistic flowering are often ranked in the same way. The Silver Age is good – in Russian literature, it's Mandelstam and Akhmatova and Blok – but it's not as good as the earlier Golden Age of Pushkin and Tolstoy and Dostoevsky. Silver played a supporting role in Athens's Golden Age of Socrates, Plato and Sophocles – the great ferment of thought and art and architecture was enabled in large part by the silver mines of Laurium on the Attica Peninsula.

Spain's Golden Age, which produced Velázquez and Murillo and El Greco and Cervantes, was also brought into being by silver. It began in the late fifteenth century with the first landing of European ships in the Americas. It accelerated rapidly, and if there was one piece of providence that propelled Spain's sudden good fortune, it was the discovery in 1545 of a mountain that appeared to be entirely made out of silver.

A stubby old Andean volcano nearly five thousand metres high,

Cerro Rico de Potosí was in Peru but is now in Bolivia. In the early sixteenth century, its slopes were just treeless tundra, too high for human habitation. It is disputed how silver was first uncovered there, but all the stories in some way credit a llama. The mountain turned out to be crazily criss-crossed with seams of silver, glinting surface-stripes that the Europeans knew at once would change their lives. They didn't know that it would also transform the world.

A Spanish settlement was established there in 1547, and the mountain's lodes justified the wildest hopes of the settlers. Within decades Potosí had become the single largest producer of silver on earth, fount of a magical elixir that was beginning to flow around the planet, a stream that swelled decade by decade until it was a torrent that kept many bobbing and buoyant (and drowned others). Over the next several hundred years, it spread its waters further and further until there is now nowhere on the planet that remains untouched by its enabling effects. It is estimated that between the sixteenth and eighteenth centuries 80% of the world's silver came from Potosí's mines.

It all began small-scale. Local workers, *huayradores*, came up from the lowlands and they worked on commission and many did well. The silver ore was close to the surface, rich and friable and easily smelted in small furnaces. In the absence of wood, the fires were fuelled by llama dung. As the sun dropped, katabic winds swept down the slopes of Potosí and the flames flared in each of the small muffle-furnaces and the mountainside flickered in the darkness. It looked sometimes as if the volcano was erupting again. Long mule trains carried silver ingots to the coast. It took them ten weeks just to get to Lima.

Meanwhile, in China the Ming government had abolished the collecting of tax in rice and replaced it with silver. At first, the silver came from Japan but when that ran out, they appealed to the Spanish. At Potosí, production increased. Prices held; demand and supply both appeared limitless. The mountain's mines grew hungry for manpower. The Spanish now imposed a forced labour scheme, mobilising indigenous people from an area of eighty thousand square miles. Climbing raw-hide ladders, with candles attached to their foreheads, the native South Americans lugged sacks of ore up from inside the mountain. Many died or were crippled in falls. Others succumbed to pneumonia in the cold thin air. More were poisoned by the fumes of mercury used in processing; suicide was widespread. Hundreds of thousands died. Silver left the town in huge quantities, and luxuries for the European settlers came back – books from the print-shops of Seville and Salamanca and silks and spices from China and Indonesia, slaves from Africa.

The town of Potosí swelled month by month, year by year. By the early seventeenth century, it was the size of Paris or London. The world's highest town had also become one of its biggest. One historian has likened its growth to settling twenty million people today in an empty spot in Alaska.

Accounts of Potosí and its riches exerted a fascination around the globe. There was something mythical about the money-mountain and its metal wealth. Written accounts of it spread. Drawings and wood-blocks appeared in published accounts of the Americas, the Ottoman manuscript *Tarih-i Hind-i Garbi* and a 1602 Chinese map of the world. Each presents a version of the same image – a bustling town dominated by a mountain, a hump of plenty, burrowed into by miners, yielding its treasure to an ever-faster carousel of commerce.

What came from the mountain was not a functional substance like iron or tin or copper. It was something more abstract, more powerful, something that would shape the centuries to come. Silver was promise and the mountain at Potosí offered it in spades –

a natural fund for the satisfaction of all known desires. It crossed the Atlantic in fleets of up to fifty vessels, with gunship escorts. A single galleon might carry two million pieces of eight – *peso de ocho*, or just '*peso*'. Spain grew briefly to be Europe's richest country. Rarely had a kingdom's fortunes risen so fast. Pesos flowed into Cádiz and then out around the continent, exchanged for commodities and finished products from France and England and the Netherlands. In turn, much of that silver left for Asia to bring back to Europe silk and spices and porcelain.

Potosí did not create the market for silver but fed an already growing appetite. At the time of the discovery of the Americas, the price of silver in Europe was already rising. Economic activity was beginning to accelerate. Until the mid-sixteenth century, most silver came from mines in Germany and central Europe – particularly those of the Harz Mountains and the Erzgebirge (the 'ore-bearing mountains').

In the late fifteenth and early sixteenth century, silver production in central Europe increased fivefold. It continued to grow until it met the tide of silver from the New World, which soon became a flood. In the 1540s, about seven hundred ounces of silver arrived each year from across the Atlantic – by the 1590s, that had soared to an annual ten million ounces.

The sudden influx had a strange effect. The prices of goods had always stayed pretty much the same, like features in the landscape. Now they started to leave their allotted places. An increase in prices of 1% or 1.5% per year is not large by modern standards but its unfamiliar nature meant it soon became known as the 'Price Revolution' ('inflation' is a more recent term). Those on fixed incomes suffered, as they always do, but merchants were able to profit from raising prices.

At the same time, the royal courts of Europe learnt to cheat at the coinage game, a game built on trust. They reduced the silver content of coins. In 1544, Henry VIII did it with the 'Great Debasement'. The French kings devalued the currency nine times in the sixteenth century. In Germany it led to a monetary crisis known as the *Kipper- und Wipperzeit* – the 'Tipper and See-saw Time' – when children were said to have played in the street with silver coins, so numerous were they. Copernicus studied the earthly movement of silver as he did the stars, concluding that it was a great mistake 'when a ruler tries to make a profit from the minting of coins by introducing and circulating new coins'.

A fidgetiness spread around the continent. Its effects were seen in price increases, in growing wealth and industry, in the impulse to produce and innovate, to search for new materials, to experiment with natural forces. The same spirit began to spread around the world, and silver was the stimulant that set it off. In the *Journal of World History*, economic historians Dennis O. Flynn and Arturo Giráldez published a piece entitled *Born with a Silver Spoon: The Origin of World Trade in 1571*. That was the year the Spanish set up shop in Manila to supply silver to China from across the Pacific – and soon, for the first time in history, all four inhabited continents were freely exchanging goods. One thing was behind it: 'The singular product most responsible for the birth of world trade was silver.'

With it came a quickening and loosening of thought. In Europe, the old hegemony of the Catholic Church had been challenged first by the Renaissance humanists, then by the Protestants. Adam Smith explained the Reformation in economic terms – 'State-supported religious monopolies behave inefficiently in many ways, thereby opening up the possibility of entry by more efficient competitors.' The Protestant Church offered fluidity, a free market in religious expression – and it found fertile ground in the silver-mining regions of Germany.

Mining, minting and Lutheranism were a productive combination in the German-speaking regions of the Holy Roman Empire. Many German towns and principalities issued their own coins. The silver came in large part from the mines of the Harz Mountains and the Erzgebirge, and the coins acted as tender in the markets of the Baltic, Bohemia, Hungary and Bavaria. In their own way, these mints helped set in motion the new age of silver, which spread across Europe to the Atlantic, across the ocean to Potosí, and from there to the Pacific and China, to connect in coming years – throughout the coming decades and centuries – all the disparate peoples of the world.

From Nebra I took a bus north-west to Stolberg and in the late afternoon watched a dark line of uplands rise from the parched plain. The Harz Mountains have one of the richest deposits of metalliferous ores in the world, and certainly one of the most influential.

At their southern end is the small town of Stolberg. It's an absurdly pretty place – all half-timbering and wonky walls and upper storeys jutting out over cobbled streets. In places, the figures of witches had been hung from the buildings, cultish emblems of the region's pagan links. The town was built where two steep-sided valleys meet and flow together down a third. Recalling early visits, Martin Luther likened the place to an eagle, with the two rivers its wings. Luther's father, who came to Stolberg to sell supplies for the silver mines, was an ex-miner himself. He often came with his son, hoping he'd take on the business. But Hans Luther was disappointed. Martin chose another path.

In his analogy, the eagle's head is the castle. It was enormous. It would have looked large even in a place five times Stolberg's size. At one point in the late Middle Ages, Germany had about 1,500 mints, turning out specie from locally mined silver. Minting made fortunes for a few, and in Stolberg, where the silver was mined in the hills above, the princely family of Stolberg-Stolberg were the beneficiaries. The walls of Schloß Stolberg swelled until it became gigantic, a stone palace built on metal wealth.

'Listen!'

Prince Jost-Christian Stolberg-Stolberg of Stolberg stopped suddenly on the steps and raised his hand. The midday bells were just beginning. A look of pleasure crossed his elderly face. 'Oh, I love that sound.'

We waited while the twelve peals rang out. Jost stood spruce in a loden jacket edged with green velvet and fronted with buttons of antler chiton. Eight hundred years of family history connected him to the bells – yet for him they were tinged with yearning. He had spent most of his life away from the town.

Born in the castle, Jost had passed his early years bicycling down its corridors, netting frogs in the ornamental ponds. Having survived the war, his family watched the Russian liberators arrive in 1945. They knew what 'liberation' would mean for the likes of them. They fled. For the next half-century, the town was in the German Democratic Republic while the family of Stolberg-Stolberg were exiled in West Germany. After reunification, the state did not return the castle to the family. But they did allow Prince Jost and his wife, Princess Sylviane, to renovate a townhouse just below it. Now they divided their time between here and Brussels.

It was Saturday and the town was full. The crowds seemed to flow past Jost's patrician figure as we walked and talked. He had a gentle, slightly other-worldly manner that I warmed to at once.

'Really, it was touch and go. The Russians had already arrived in this sector. Other families had begun to disappear. A British contingent got through on 8 July – they'd come to rescue us. We had just hours to get whatever we could. I remember everyone madly rushing around. I desperately wanted to take my bicycle. On the edge of town the Russians blocked the road – the British jeeps went bumper to bumper with them. Touch and go . . . Thank god the Russians gave way. Otherwise it would have been Siberia or . . . well, it hardly bears thinking about. Of course, I was not aware of all that. I was just pleased to have my bicycle.'

In Stolberg, the legacy of silver flashed dimly from centuries past. The last ores had been scraped from the mines in the late nine-teenth century, but by then the peak had long passed. At the Alte Münze, the 'old mint', the town's history had been preserved in a collection of tools and coin-presses and rolling-mills.

'Most of this was collected by my father,' Jost said as we walked through the exhibition rooms. 'Before the war he realised that all the old minting machines were collecting dust in attics all over the town. No-one cared much at that time.'

Yet here they all were – celebrated and re-presented with tasteful lighting and explanatory texts, sanitising for a contemporary audi-ence the real tasks, the real lives of those who transformed the dust of the earth to currency.

'Look,' Jost was bending down to read a caption. 'It says here they worked from five in the morning until seven at night. What a life!'

Back at his house, everything was supremely elegant. Jost and Sylviane had condensed the castle's spirit into these few, finely proportioned rooms while the castle above was being renovated as a luxury hotel. Sylviane sat reading in a Louis Quatorze chair while Jost brought from his library a velvet-lined album.

'I have only a few left.' Standing in the light of a tall window, he showed me his coins. 'This is a thaler – sixteenth century. Thirty grams. Is it . . . ? I cannot read it well . . .' He took an eyeglass from his pocket. 'Ah! 1598.'

On the coin was stamped the Stolberg stag. The same heraldic beast formed the buckle on his belt.

He picked others – later ones, from more recent centuries, and then another, much more delicate. 'This is from the thirteenth century. Just 0.37 of a gram.'

I held it in the palm of my hand. It felt as weightless as a leaf. Yet these things built armies, launched ships and dug mines. They shifted goods and resources from one place to another, stirring desires in people that they didn't even know they had.

In the winter of 1777–8, Johann Wolfgang von Goethe visited the Harz Mountains for the first time. He was twenty-eight and it turned into a journey of revelation. Silver and silver mines brought him to the Harz (he was Commissioner of Mines of Saxe-Weimar), but it was rocks more generally that he took away. The Harz sparked in Goethe a fascination for geology that became the basis for wider enquiries into the dynamics of change and the nature of the earth – questions that occupied him for the rest of his life and became of equal importance to him as his literary work.

Goethe was born and grew up in Frankfurt. In his early twenties he already cut something of a dash in the city, knocking out lyrical verses with great speed and apparent ease, speaking on any manner of subject with an almost hypnotic fluency. Sitting beneath a tree, lying in the grass, standing in a public square, his garrulous presence drew crowds. When he was out walking, women and children flocked to him. He was thought by many to be a prophet, a genius, even some sort of Jesus figure. At the age of twenty-five, he published his novel *The Sorrows of Young Werther* and its tale of a hypersensitive youth propelled his celebrity far beyond Frankfurt. Considered the world's first bestseller, the book went into seven printings in its first year. It was translated into most European languages. Napoleon later told Goethe he read it seven times. The great salons of Europe were buzzing with his name – and that was the moment he chose to turn his back on them.

At the age of twenty-six, he abandoned Frankfurt for the tiny, impoverished duchy of Weimar. The Duke of Weimar had invited him to help restore the territory's fortunes and, almost on a whim,

Goethe agreed. The duke – who was even younger than Goethe – made him a privy councillor and also put him in charge of mines. The first year was a riot. The two men spent their time carousing and drinking and visiting taverns. Then Goethe became serious. The republic needed money and he immersed himself in the attempt to reopen Weimar's flooded silver mines. As research, he took himself off to the world centre of mining. In November 1777, he went to the Harz.

It was an impulsive decision, as his departure from Frankfurt had been. He told no-one he was going, neither the duke nor his friends. As he rode out of Weimar, icy winds kicked up clouds of newly fallen snow. On the second day, the skies cleared, the temperature dropped and in the crisp light he caught his first sight of the Harz and its rounded peak. The Brocken!

He travelled alone and incognito. He took the guise of a painter named Weber and as such found he could pry without suspicion into the working of the Harz mines. He quizzed mine captains and climbed down shafts to witness the workings up close. He had a near escape from a rockfall, blithely thanking 'fate' for his survival. Worlds and possibilities opened up.

The mines of the Upper Harz provided almost half of Germany's silver. Not only did they produce coins for commerce, but they acted as a laboratory, a proving ground for technology in that early scientific period when big ideas and practical solutions freely mixed. Christopher Polhem was known as the 'Archimedes of the North' and following a visit to the Harz from Sweden in 1727, he invented a 'Sipho-Machine' based on vacuums and air pressure. The polymath Gottfried Wilhelm Leibniz (known as the 'last universal genius') spent several years in the Upper Harz. He had seen what wind could do in draining the Dutch polders and proposed a series of windmills – including a horizontal windmill that could, he assured his backers, operate in the slightest breeze. It didn't catch on. But Leibniz's thinking was a great influence on Goethe's own, in particular his emphasis on unity and process and the overarching notion that everything in the universe is in some way *beseelt*, or 'animate'.

In December 1777, young Goethe arrived in the new mining town of Clausthal. In its wide streets and handsome wooden-clad houses, he sensed an ebullient prosperity. It was larger than the city of Weimar. In his diary he wrote of 'a region where the mining towns grow happily, based on the benedictions of the underground'.

A part of Goethe was always an optimist. His lifetime – from the mid-eighteenth century until the 1830s – coincided with Europe's great industrial leap forward. Though mindful of the costs, he welcomed progress – the adventure and the promise. In his last years, he became hugely excited by grand schemes like the Panama and Suez canals, and another that would connect the Rhine and the Danube. 'The resulting benefits for the entire human race,' he believed, 'would be incalculable. . . I'd love to live to see it; but I won't.'

On that first visit to the Harz, in his late twenties, he admired the ingenuity of water management, the networks of cascading water-wheels below ground, the flatrod and suction pumps. He saw the immense efforts and triumphs of extraction. He watched barrels of ore rising from the shafts, and he saw the same ore travelling as rubble on flat-bed carts, to be changed to silver. All things for Goethe were connected. In the ore, and in the growing towns, he saw something that was as relevant to him personally as it was to the great technological apparatus around him – transformation, development, becoming.

*

From Stolberg, I travelled north to the Upper Harz and for several days poked about the old mining districts. In Clausthal, I met an industrial archaeologist with a wonderful enthusiasm for the old mines who said, with a flourish: 'The modern world, it was made here in the Harz.' I sat in small cafés where old men with enormous moustaches ate bowls of soup and stared into the middle distance. The last mine closed in the 1980s. Tourism was now the thing – small mining museums with proud signboards and the UNESCO logo.

One morning, I hiked up into the forest. The air was sharp, sweetened by the smell of spruce sap. The trees had been clear-felled and everywhere were brush piles and woody detritus. From somewhere far off, a woodpecker was rattling out a message. I left the loggers' track for a small green path that led away around the hill beside a mill stream. There was something nameless and irresistible about that path and the stream with its border of pole-straight trees still standing and the morning sun slanting through them.

I stopped and looked more closely. The stream was tea-coloured, tinted by humus from the moors above, from peat and blanket bogs. The water was kept in the ditch by a retaining wall of dolerite. I knelt down to inspect those walls and the squared-off, interlocking blocks. I thought: *These all had to be quarried; they all had to be dressed and carted here and put in place and the cutting had to be cut.* Three hundred miles of these ditches run through the mountains. They were *Kunstgräben*, fuel pipes for the mines' waterwheels. Numerous reservoirs and ponds had been built to give the water a head. In the walls was a diligence that appeared to go far beyond what was necessary.

I came to a small dam. Above it, the stream tumbled over a slab of granite and into the pool. Clear ripples of water pulsed across the granite. From the pool, a sluice sent the stream one of two ways. Half the flow carried on down the slope, fast and wild through its rough-edged gulch. The other half was siphoned off into the *Kunstgrab*, where at a consistent sixty litres per second, on a consistent 1–2% slope, it began its orderly way to work.

Nature one way, industry the other. Back at the beginning the two went hand in hand. Nature was generous. The silver ore was close to the surface and the forests provided timber for smelting. As the years passed, the miners had to travel further for the wood, dig deeper for the ore. In the fourteenth century came the plague, which helped put the brakes on Europe's economy. The value of silver collapsed. But it was temporary. Activity surged again from the late fifteenth century, years of plenty when Harz silver played its part in the beginnings of global commerce. Here at the source were unintended consequences. The forests grew thinner, the miners ended their short lives coughing, and if the wind was in a certain direction cattle would drop dead from lead fumes as they grazed. The rewards for the houses of Brunswick-Wolfenbüttel and Celle-Calenberg, the Kurfürst of Hannover (and the English king) and later the Prussians were enough to keep it all going.

Goethe might have been thrilled by engineering projects but in his version of *Faust*, he presents a character who is ultimately destroyed by them. Faust is all of us, yearning constantly for more. He has drained the swamp, freed up good land, but still he is not satisfied: 'The worst torment / To have so much, yet still to want!'

One thing is driving Faust crazy: a single piece of undeveloped land with a grove of linden trees, a chapel and an ancient cottage. He instructs his men to relocate the cottage's elderly residents. There's a scuffle – a fire starts, the couple are killed and the trees destroyed. The whole sorry incident – which comes at the play's climax – stands for the dangers of good intentions. Faust was only trying to improve the lot of his people. During the eighteenth century, improvement was the noble aim of all early stages of technology, the hoped-for result of trade and commerce. It was impossible to imagine then that it was having wider impacts on the planet.

Now we have evidence. Cores taken from Arctic ice, for instance, have recently revealed layers of atmospheric lead pollution so distinct they can be matched to historical moments. Lead was a by-product both of silver extraction and its smelting. A recent report in *Proceedings of the National Academy of Sciences* presents a precise chronology of lead levels. There was a surge as the Romans introduced the silver denarius, a dip when the empire collapsed, and a spike again as rich lodes of silver were discovered in the Harz in 968. When famine and pestilence forced an abandonment of the Harz mines between 1002 and 1016, there appears 'an immediate and persistent decline in the record'. Another spike came a few hundred years later: 'The increase between the mid-15th century and the mid-16th century almost certainly reflected the revival of silver-mining in the Erzgebirge and Harz Mountains.'

I set off back down the mountain, through the cleared forest. Snapped-off trunks lay like broken limbs on a battlefield. This was not cyclical logging. It was remedial. The Harz forests are sick, and mining is partly to blame. Centuries ago, digging and processing of ores had stripped the Harz of its native beech and oak forest. The forests were re-stocked with spruce, which grew fast and straight and was good for pit-props and pump-rods and for making charcoal. But it was a monoculture, and monocultures are always vulnerable. An infestation of the *Borkenkäfer* beetle spread through the trees. One outbreak – the *Große Wurmtrocknis* – destroyed thirty thousand hectares of forest. In November 1800, a vicious storm struck what remained of the weakened trees and flattened them.

From the 1930s, spruce was again planted, this time in even larger numbers. The teams who did the work were celebrated by the Nazis as heroes of the Reich. All those trees are now the same age, and recent hot dry summers have affected them too. The *Borkenkäfer* beetle is back and it's thriving. The impression of travelling through the Upper Harz – which works so hard to celebrate its mining past and the ingenuity of extraction – is of a landscape devastated by an unseen horde.

At Rammelsberg, I was given the address of an elderly wood-worker who made mine-models up in the village of Altenau. His shop looked closed. But when I pressed down the handle, the door creaked open. There was no-one about. In the shadows I could make out shelves of wooden figures and wooden chandeliers and scores of wooden shapes I couldn't identify.

A fluorescent tube flickered in an adjacent room, and came on. Still no sign of life. I stepped into the lit-up space. One entire side was taken up with a model – *Silberbergwerk*, 'silver mine': an intricate tableau divided into above ground (trees and hills and various works) and below ground (cross-section of a mine). At one end was early mining, searching for ore with dowsing rods and cracking the bedrock by fire-setting. In the next scene were the smelting works and the red glow of a furnace.

Then there was a whirring and the model started to move. Two men on the surface were working a winch; from down in the shaft, a kibble began to rise. A horse was traipsing round its whim and stamps were crushing the ore; far below, at the foot of a long ladder, a team were hammering at the rock.

I was aware of a man in the doorway behind me.

'*Wunderschön!*' I said.

Did he smile? He might have nodded faintly. 'My grandfather's work.'

He shuffled off again. More lights came on through another doorway. Two more tableaux clunked into life before he reappeared. 'This – Altenau in Winter.' He raised an arm at a snowy scene, with a green summer moment in the Erzgebirge at the far end.

'*Ja, ja, es ist alles hier,*' he said.

His name was Hermann Meier. For over two centuries, from the time of Goethe, his family had been making folk figures, saintly statues and household utensils from wood. And not just the men. Hermann had four aunts and they were all woodcarvers; his daughter also carved. But it was his grandfather Richard who raised the art to a new level with these models. In the 1920s he took his creations on tour – first on a handcart to Christmas fairs, then, as their popularity grew, on a larger wagon, driving around Nazi Germany and dropping the sides of the specially adapted trailer. After the war, he received commissions to create models of state projects, dams and high-rise buildings.

'He began painting when just a child.' Hermann's voice was cracked. 'Carving too.'

At twenty, Richard had a serious accident. He spent six months in hospital, yet even in bandages, he worked.

'Creating was what saved him. That's what he always said.'

The following year he had another accident, more serious. He was working in a paper factory and fell between two boiling metal vats. They thought he was dead, but he survived – with severe injuries to his face and his working arm permanently paralysed. He spent a year and a half in hospital and learnt to work with his left arm. 'Creating these was what saved him.'

I bent to look at the figures, at the adits down below pushing into narrow spaces, the details I'd missed the first time, the miners' clothes, the tiny tools. These models were devotions. In Susan Stewart's classic study of miniaturisation *On Longing,* she speaks of the appeal of mechanical models as 'a transcendence which erases the productive possibilities'. The model is aesthetic, where the original is merely functional. Richard Meier's scene does not present the 'why', the demand and use of the extracted metal, so it makes the labours of silver-mining look like some strange abstract quest.

It was Hermann Meier who told me of some old silver pits outside Altenau. The way there led down one steep-sided valley, and up another. Where the two valleys met was a dam and a wishbone-shaped reservoir. On its grey waters cruised a large pleasure boat, looking like a big beast pacing a small cage.

The pits were in the forest. I reached a house where four people were sitting on an upstairs balcony, smoking in silence. One of them stood and pointed at a path that curved into the trees: '*Ja, ja*. You find old ones there.' She then returned to the business of smoking.

Up a side valley and a couple of false trails and some scrambling, and then amidst the trees the scar of old mineral workings. These were not mines but scourings, open pits with mounds of spoil and scant vegetation around them. It was how it started, where silver coins began in ore, were then conceived in the heat of charcoal furnaces, moulded into ingots, then pressed into service in a thousand mints. Inside the pits, soil had accumulated and branches and small trunks were spillikined against the sides. There was a line of similar pits running up along the slope where the miners had chased the lode. The silver had run out pretty quickly.

I sat on one of the spoil-heaps and looked across the valley. A large open patch showed where the beetle-gnawed spruce had been and birch scrub had started to fill in. I watched the clouds overhead, quick mountain clouds with a steely sharpness at their edges, and the deep blue beyond. The wind soughed in the trees and I lay on one elbow to examine the spoil close up, its whitish sands and greywacke rubble and hornfels.

A worker ant was labouring across them. It had a crumb of something in its mandibles. I became absorbed by its progress, focusing on its steps – up, down, over and beneath – watching the way its thorax articulated as it went, bending to the micro-mountains, where each of its tiny legs scrabbled for grip. The sun burst from a cloud and its sudden brightness crossed the ground, and it caught the milky quartz and glowed on the back of the ant. I was aware of something familiar. Not happiness or pleasure, but joy – that flying visitor, affirmer of life and a wider belonging in the world. Triggered by what, exactly? By sun and forest, by the broad cloth of the day, by a pile of rocks, by an ant.

Goethe first came to the Harz to find solutions to the problems of Weimar's mines. What he found was something that shaped the rest of his life, something transcendent. What he found was joy.

'Father of love,' he wrote in *Winter Journey in the Harz*,

> If from your song book
> There sounds one note his ear can hear
> Refresh with it his heart!
> Open his clouded gaze
> To the thousand fountainheads.

During those weeks in the Harz Mountains, Goethe duly found his heart refreshed and his gaze opened to countless springs of joy. The single note that he heard from nature's song book, which replenished him, came from rocks. His already deep devotion to the natural world was given new force by the Harz Mountains' diverse geology. In the silver pits he found himself intrigued by the practicalities of mining, but the exposed cliffs and the granite of the

Brocken stirred bigger thoughts – about origins, how mountains are formed, the clues to the primal events that lie at our feet, the possibility that solid rock might reveal the greatest of all mysteries – how things change from one thing to another.

He began collecting stones. He was not alone. In Europe at that time there was already a healthy market in crystals and fossils. The philosopher Herder, Goethe's friend, commented tartly: 'At that time the person was nothing, the stone everything . . . Everyone mineralogized.' As a collector, Goethe avoided aesthetics. He sought samples that might provide evidence for his geological ideas and speculations. By the time of his death in 1832, he had one of the largest private rock collections in Europe.

Goethe's thinking was driven, above all, by the idea of connections. Scientific enquiry was only beginning to be divided into disciplines and his detailed musings and experiments roamed freely around physiognomy, botany, galvanism, magnetism, geology, meteorology and optics. Where others specialised, he followed his own expansive curiosity. 'Goethe, though devoted to science,' wrote the Nobel laureate Charles Sherrington, 'had not at root the scientific temperament [and] lacked the sublime detachment of the scientific thinkers.'

Goethe brought to science a poet's sensibility, a genius for metaphor that was concerned more with likeness than with causality. Such a view owed something to an early immersion in alchemy and the Hermetic tradition. At the age of nineteen, by his own admission, alchemical texts occupied his reading and though much of the language was 'obscure and incomprehensible', there was something about it that sustained his attention. He studied chemistry and conducted experiments, describing it all as his 'secret love'. He set off in pursuit of a material known as Virgin Earth. Looking back many years later, he recalled that the substance he produced was quite useless but the process of making it was 'beautiful'.

To read Goethe on science is to experience an immense freedom of thought, a wild frisson of imaginative association. He was wrong about many things – his theory of colours, much of his botanical work and his geology have all long since been disproved – but he also had successes. He stumbled on a number of ideas that later became scientific orthodoxy. He was one of the first to write of an 'ice age' – explaining erratic boulders of granite in northern Germany by suggesting that the area was once covered by slowly moving ice, which then melted. He identified the intermaxillary bone in the jaws of humans; its apparent absence had hitherto been evidence of humanity's distinction from the rest of the animal kingdom (Goethe was very happy to dispel the notion of anthro-exceptionalism). And Charles Darwin identified Goethe as one of the few thinkers who, decades before his own *On the Origin of Species*, understood the overall principles of evolution.

Goethe became increasingly convinced that fathoming the natural world came not from identifying species and substances – the distinguishing of one thing from another – but from what links them. 'It is evident,' he wrote, 'that all things in nature have a clear relationship to one another. Every creature is only a tone, a shade of one great harmony.' From the early years of the industrial age, his words reach our own with an urgent clarity.

Professor David Seamon, who wrote of Goethe's phenomenological approach, described the effect of his work as a '"folding over" of natural phenomena so that things unjoined before now connect in relationship'. Goethe was less a scientist than a proto-ecologist – not least because he wrote of the natural world with unashamed rapture. 'The world is so great and rich,' he said, 'and life so full of variety, that you can never want occasion for poems.' Goethe recorded his excitement as he came closer to a grand overarching truth. 'I am beginning to grow aware of the essential form with which, as it were, Nature always plays . . . If only I could communicate the insight and joy to someone.'

His studies led him far from conventional Christianity but not towards the mechanistic view of Descartes and Newton. Instead he bobbed about in a sea of pantheistic reverence:

> I worship him who has infused into the world such a power of production that, when only a millionth part of it comes out into life, the world swarms with creatures to such a degree that war, pestilence, fire and water cannot prevail against them. That is *my* god!

Revelatory joy and unity were twin drivers of Goethe's work. In his poem entitled simply *Freuden*, 'joy', he watches a dragonfly flitting back and forth over a spring: 'chameleon-like . . . now blue, now red / And now of greenish hue.' By coincidence, William Blake's famous poem about joy – *Eternity* – was also prompted by watching a creature on the wing, in his case a butterfly: 'He who binds to himself a joy / Does the winged life destroy / He who kisses the joy as it flies / Lives in eternity's sunrise.'

Fleeting joy again. It is there in Faust's dying cry: '*Verweile doch, du bist so schön!*' (Linger awhile, you are so beautiful!). But Goethe here is making a more subtle point. He follows the dragonfly in flight and when it lands, he is at last able to examine it closely. In repose, the creature's multicoloured appeal has vanished:

''Tis of a sad and dingy blue –
Such, Joy-Dissector, is thy view!'

Chopping the world up into discrete chunks, dissecting it, was a growing tendency in eighteenth-century Europe. Examining details may reveal certain traits, but it misses something too. Goethe's early interest in alchemy left him with notions that swam against the rising tide of the Scientific Revolution. He never lost his belief in the patterns that flowed between disciplines and was constantly struck by the beauty of such connections. Joy comes from a brief glimpse of unity; true understanding lies in the whole.

I hadn't planned it that way. A news item, caught by chance the night before, had spoken of a total lunar eclipse over Saxony. When I woke just before dawn, on the edge of the forest below Torfhaus, I watched from my sleeping bag the rose-red disc of the moon go through a week's waning in a few minutes. It was still shrinking when it dropped behind the trees on the skyline. I was due to climb the sacred mountain of the Brocken and the eclipse left in its wake an auspiciousness that made me smile. It also left a faint knot behind my brow that was there all morning, a warning of cosmic frailty.

The Brocken is not a high mountain by any means but it's one of those places that dominates its surroundings. Head north and you'll hit nothing higher until Norway; head east and it's the Urals. To the south-east is Mittelberg, where they'd buried the Nebra Sky Disc and watched the solstice sun rise above the distant peak. Witches congregated in the Brocken's mists to frolic with the devil. Ghouls hid in its rock-shadows. The Stasi used it as a listening post.

I set off towards it, following a flattish path through a peat bog.

Goethe's first visit to the Harz Mountains may have been professional, but as he went down more and more mines, a personal urge took over. He wanted to climb the Brocken. On a snowy morning in December, he rode up from Altenau to Torfhaus. He asked a forester about going up the Brocken. Impossible! Look at the fog. Anyway, the forester said, he'd never even tried it, didn't see the point.

Suddenly the fog cleared and the peak was, recalled Goethe, 'as clear as my face in the mirror'. He persuaded the forester: 'I scratched a sign into the window pane as a testimony to my tears of joy.' The two of them set off on horseback. It was to be the first ever recorded ascent in winter, and for Goethe the journey helped elevate those weeks in the Harz to the status of private myth. He later described the mountain as 'an altar of sweetest thanks, / Which an ancient people's presentiments / Bedecked with hosts of spirits.'

Coming out of the shade of the Königsberg peak, I blinked in the early sun. It lit up the trail ahead – bilberry leaves beside it, gossamer strands and buzzing clouds of bog midges. I sat for a while beside a *Kunstgrab*. I watched the water; I ate a bread roll. The trees were short here; many were dead, with their sad broken trunks and snapped-off branches. I listened to a cuckoo, to its two-note woody call – *Kuckuck* is the name for a cuckoo in German. I carried on.

The forest thickened. There was an old concrete border post with a roundel of hammer and dividers and written beneath it: DEUTSCHE DEMOKRATISCHE REPUBLIK. Five hundred miles of multistranded barbed wire had run through here, a hard metal divider that was less a border than a national dissection. I crossed into the old East Germany. The cuckoo again.

As the path steepened, the trees became fewer and smaller. The boggy ground opened out. Amidst the heather were scars where it had cracked to oily black peat hag. The Brocken is the remains of an upwelling of magma that has left the soils acidic and the rock around it ripe with minerals. But at its heart, it is granite, pure granite, going down deep into the earth's mantle. In places the granite is exposed – gritty slabs beside the path, and vast grounders in a blockfield above. I imagined a time-lapse of one frame every hundred years, and the whole slope in flux. Still the cuckoo – *kuck-uck*.

Goethe's essay 'Über den Granit' – 'On Granite' – is a lovely piece of Goethean phenomenology, an expansive blend of emotion and geology. In it he writes of the Brocken:

Here, on this primal and everlasting altar raised directly on the ground of creation . . . I feel the first and most abiding origin of our existence; I survey the world with its undulating valleys and distant abundant meadows, my soul is exalted beyond itself and above all the world, and it yearns for the heavens which are so near.

For Goethe, granite was the first solid substance of the earth, the ur-rock – the 'oldest, firmest, deepest, most unshakeable son of nature'. It stood in direct contrast to his other main subject: 'the human heart, the youngest, most diverse, most fluid, most changeable, most vulnerable part of creation'. By studying one, he thought, he could understand the other, because in nature all things are analogous.

Quartz, feldspar, mica. The three constituent parts of granite came, according to Goethe, at the beginning. They are combined in a particular way – the same way that nature as a whole combines its parts: its trees and flowers and rocks. Granite 'neither contains its parts, nor is contained by them, but possesses a complete integration, a trinity'. He speaks of *die Dreieinheit*, 'the three-in-one', and implies in doing so the triune structure that pervades the universe (his claim is supported by the more recent discovery of subatomic particles of neutron, proton, electron). If any of granite's elements should become dominant, or 'step out of itself', then the essential harmony is disrupted and 'a state of anarchy' ensues. In such a state, other minerals are formed, including metals.

Two hundred and fifty years of geological research, fuelled by the enormous rewards of mineral extraction, has shunted Goethe's theories into a siding. He was wrong: there is no such thing as a 'first rock'. In 'Über den Granit' he admits that he is not best placed to pursue the study of granite, but he'll do it anyway: 'my efforts will afford others the opportunity to go further'. Goethe got the details wrong – but he sensed the whole.

Nature has long been the house magazine of academic science, a global multidisciplinary organ with unparalleled prestige and citation value. The first article in the first issue in 1869 was written by the great biologist T. H. Huxley: 'When my friend, the Editor of *Nature*, asked me to write an opening article for his first number, there came into my mind this wonderful rhapsody on "Nature", which has been a delight to me from my youth up.' The quote, he explained, was from Goethe:

> Nature! We are surrounded and embraced by her: powerless to separate ourselves from her, and powerless to penetrate beyond her . . . She herself is tireless. Her present is eternity. Always is she a whole and yet never is she complete. All the time she constructs and she destroys all the time. Life is her fairest invention and death is her device for getting more life.

Huxley then reflected on Goethe's 'rhapsody' and its relevance for *Nature*'s first issue: 'Long after the theories of the philosophers whose achievements are recorded in these pages, are obsolete, that vision of the poet will remain as a truthful and efficient symbol of the wonder and mystery of Nature.'

*

It was still early when I reached the Brocken summit. The early sun sliced the exposed rock into bright surface and sharp shadow. Out of it rose a bulbous-domed building from the Cold War. Nicknamed the 'Stasi Mosque', it bore the official title of 'Urian' after one of the spirits of the mountain. Each morning inside that dome, twenty-eight officers would put on headphones and twiddle dials to tune into the police and military radios of the enemy, their compatriots in West Germany. A few weeks after the collapse of the Berlin Wall, a crowd marched up here and demanded their mountain back. The guards knew the way history was going; they let them in. The crowd was said to have numbered one hundred thousand people.

I left the summit in good spirits, following the long curves of the path. The day was scribbled overhead in signatures of high cirrus. Beneath it, the plains of central Europe stretched out in hazy miniature, with tiny towns and postage-stamp fields. Down towards the treeline, the path steepened to a chaotic stairway of granite boulders. Descending it was an exercise in focus. Eyes had to work ahead of feet, seeking each landing point in the moment before it was needed. It seemed to work only at speed, and soon I'd reached the treeline and entered the forest.

Except it wasn't a forest. It was a boneyard, an apocalyptic scattering of timber. I tried to take it in. I'd grown used to seeing dead spruce, but not like this, not this dense. It looked like the aftermath of a chemical spill or radiation spike. Broken trunks, fallen trees lay in matchstick patterns, layer upon layer and all of it deathly grey, as if colour itself was life. I found it deeply shocking.

A little way on was a small outpost of Nationalpark Harz. A ranger in a wide-brimmed hat was standing outside, and his ranger's eyes were ranging across the valley and the scattered trees. 'Maybe it does not look nice – but it is the best thing.'

He was a little too quick to explain. I imagined his day peppered with visitor complaints. 'We used to clear the trees away, but now we leave them. That's our policy. *Natur Natur sein lassen*. Let nature be nature.'

A bit late for that. The entire Upper Harz was testament to not letting nature be nature. Plundered metals, bedrock turned to a honeycomb of shafts and adits, waters diverted, soils poisoned with lead, native forest twice stripped and replaced with an arboreal monoculture which is now prey to beetles and weakened by more and more frequent droughts.

'Please.' He led me up over the lifeless logs to a small clearing. In its disorder was a nascent diversity. Tiny seedlings were appearing – crowberry and rowan and self-seeded birch and willow. 'It is recovering.'

Natur Natur sein lassen – 'let nature be nature', like 'let bygones be bygones' or 'let kids be kids'. Don't intervene. Children, the past, nature – these things all have their internal dynamics, and we try and shape them at our peril.

A few hours south by train, to the small town of Weimar. In Frauenplan Square, I approached the townhouse which Goethe had shared with Christiane Vulpius. His domestic arrangements were a scandal not just because the two were not married but because she came from the lower classes. They had seven children. Only one survived.

Goethe lived here for nearly fifty years. 'He missed the west,' said Thomas, the curator. We were in the garden, beside an espaliered vine planted by Goethe. Thomas cupped a cluster of pea-sized grapes in his palm. 'He wanted to grow grapes – to remind him of a warmer place. But really it is too cold here. In Frankfurt, maybe. Not here in Weimar.'

In the far corner of the garden stood a squarish pavilion. Thomas fiddled with some keys before finding the right one, and he pushed at the door. It stuck for a moment, then scraped open. The air inside was stale and cold.

'Sorry about the mess.'

'I don't mind mess.' After the museum sterility of the main house, the pavilion felt like hidden treasure. We creaked up the wooden steps to the first floor. A tripod stood with its camera removed, and there were vague signs of cataloguing. The whole place had the dusty sense of long abandonment.

'No-one's been here for a while,' said Thomas, absently.

There were rocks everywhere – on shelves, on windowsills, in the spaces beneath tables; and not just minerals, but fossils, petrified wood, the tooth of a rhinoceros, the jawbone of a stag.

'These cupboards he had made specially for his minerals.' Thomas pulled open the drawers of several chalky-grey cabinets. Inside were more rocks, crammed together in little card boxes. Some had labels in Goethe's own copperplate hand, and there were source notes, catalogue numbers in ink, classification signs – some just a single Greek letter. In places there was a flash of mica, a seam of milky quartz as the drawers slid out. But this was not a collection to win prizes at mineral shows. Arranged in 'suites' – *Gebirgsarten* – the specimens were in large part brown and lumpy.

As well as geology, Goethe did extensive work on botany. Plants are full of poetic appeal – the beauty of a leaf curled up in the bud, all of what it would become, the symmetry and multitude of forms. He loved plants for these reasons, as he loved rocks, and wrote a long essay called 'The Metamorphosis of Plants' and a poem of the same name (because neither prose nor verse could quite say it all):

Every plant now declares those eternal designs that have
   shaped it,
Ever more clearly to you every flower-head can speak.

He believed that all plants stemmed, as it were, from a primal
or ideal plant, the *Urpflanze*, and the wonderful blooms of the
vegetable kingdom, the seasonal diversity, were but rungs on 'the
spiritual ladder', part of a perpetual progress towards the 'summit
of Nature'.

The charm of rocks is much less obvious. Goethe never tired of
them. Wherever he travelled, his baggage grew weighty with stones.
He promised himself not to collect any more, but he couldn't help
it. 'The moment I get near mountains, I become interested again in
rocks and minerals.' Returning from Italy and dreading the journey
back over the Alps to the drab north, he bought himself a geological
hammer to 'tap on the rocks to drive out the bitterness of death'.
In old age he is remembered for stealing off into the park with his
hammer in times of stress: 'Allow me to hurry down to my rocks
. . . it befits old Merlin to renew his friendship with the primeval
elements.'

His actual writings on geology are limited. He chose not to
publish 'Über den Granit' in his lifetime. 'Stones are mute teachers,' he wrote; 'they make the observer mute, and the best that one
learns from them cannot be passed on to others.' Goethe knew well
that the closer you get to the core of truth, the less efficient words
become.

Rocks do not do much – at least in the span of our own brief
lives. They lack the frenzy of organic life. But they play their part in
the drama of creation. They nourish all living things. Their movements are slow, tectonic creepings visible in million-year frames.
What we see on the surface is just a snapshot in geological time, just

a fraction of the lightless geo-kingdom below, where all is in flux. Goethe loved rocks not just for their process but for their 'sublime tranquillity'. Of all the earth's wonders, they are the grandest, the oldest and the slowest.

In the 1820s, as he approached his own death, Goethe grew more and more aware of the world's jeopardy. It was speeding up. Not an untypical complaint from an old man, but Goethe gave it his own spin. He coined the term *Veloziferische*, combining the Italian *velocità*, 'speed', with *Luziferische*, 'devilishness'. In canal-building, road-building and locomotion, the primary motivation was to cover ground faster, to gain more time. One thing above all enabled it: the proliferation of credit. In one of his last letters, Goethe wrote: 'Just as the steam engines are not to be stifled, neither is the new morality: the briskness of trade, the rustle of paper money, the accumulating debts to pay off other debts.'

Karl Marx noted the same thing several decades later: 'The credit system hence accelerates the material development of the productive forces and the creation of the world market', warning that such credit also generates 'crises'. Unlike Marx, Goethe did not attach the idea to any sort of political theory but rather saw it, like geology, as a cosmological force. The twentieth-century French economist Jacques Rueff called Goethe the world's 'greatest money theorist . . . He was, to be sure, no economist yet it was he who in *Faust (Part II)* clearly demonstrated that inflation was and could only be an invention of the devil.'

From the silver mints of the Harz to the mountain of Potosí, from the coffers of the East India Company to shiploads of Spanish pesos, the wheels of commerce had been spinning ever faster. As they accelerated, so the methods of exchange became more elaborate, further speeding up the system.

Silver had made barter look cumbersome, but once value had been attributed to coin, the next abstraction proved easier – paper money. From silver to paper, from paper to digital. Money now exists in cyberspace. Cryptocurrency is the purest distillation of trade, tender that has no physical existence at all. Bitcoin lacks any controlling authority, nor is it constrained by the difficulties of digging it up. So how to maintain its value? With a puzzle. Bitcoins are generated by trying to guess a sixty-four-digit hexadecimal sequence. The odds of finding a coin are about one trillion to one. Like silver, the material is there waiting for free; the challenge is to find it.

With cryptocurrencies, money has come full circle. Bitcoins are 'mined' and, like physical mining, the rewards are vast, the true costs deferred. The computer capacity needed for Bitcoin-mining is so enormous that it has led, since its inception in 2009, to the release of two hundred million metric tonnes of CO_2 – or 0.1% of the world's entire greenhouse gas emissions.

Increases in the use of silver led to the Price Revolution of the sixteenth century. Since then, inflation has become an acceptable side effect of economic growth, a necessary evil to be kept in check by the careful manipulations of central banks. But monetary inflation is not the real debt. The superfast turmoil of trade that has increasingly occupied the nations of the world has been running up a tab against nature itself. Silver has been turned on the very earth that yielded it.

Goethe's idea of *Veloziferische* has contemporary implications that, two centuries ago, even he could not have imagined. Weimar itself is now forever associated with the madness of hyperinflation. In late 1922, a loaf of bread here would have cost you 160 marks; a year later it was two hundred billion marks.

The town is now little different from any other in Germany, with its chain stores and restaurants, its gift boutiques full of coloured umbrellas and key rings and souvenir mugs. In the shop of the Goethe Museum, a swipe of a plastic card can give you a bust of Goethe, a bottle of 'historical scented water' or a pair of 'diabolical red' oven gloves printed respectively with *Faust I* and *Faust II*.

In the garden here, in the attic dustiness of his old pavilion, was something less ephemeral. Rocks and more rocks – on tables and shelves, in drawers and boxes – one man's gathering of the litter of cosmic process, an attempt to understand their slow formation and trajectory. The effect of being among them was to glimpse a stillness that, for a moment, looked like the truest state of the earth.

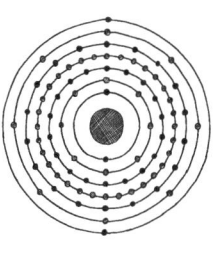

6

RADIUM

AT THE END OF THE nineteenth century, a metal was discovered that turned the known world on its head. Nature had
become more predictable of late. Each of the sciences was putting
things in order, and chemistry, the newest of the three main
branches, was proving every bit as tidy as the others. Mendeleev's
periodic table had emerged as a very efficient way of arranging the
basic components of the universe. Each element seemed to have
its own pre-booked seat at the table, arranged according to atomic
weight, and new ones simply slipped into the empty spaces. And
once they'd found their place, they stayed put. They could be built
up into compounds, the metals into alloys, but they couldn't be
broken down further or transformed. But here comes radium,
lurching and fizzing into the room like a hyperactive socialite. Not
only was it in a state of transition but it appeared to contain its own
internal power source. Everything about it was impulsive, exciting,
rebellious. This was a party now.

In H. G. Wells's 1913 novel *The World Set Free*, a professor of
physics speaks of radium as a 'fantastic exception, a mad inversion
of all that was most established and fundamental in the constitution of matter'. In a lecture at the Lowell Institute in Boston, William
James suggested that the surprise of finding radium was 'as if I

should now utter piercing shrieks and act like a maniac on this platform'.

It began in 1898, when Marie Curie was studying a lump of uranium in Paris. She noticed that its levels of radioactivity were inexplicably high. There was something else in there. She set about reducing the ore, grinding it in pestle and mortar, filtering it, precipitating it, dissolving and crystallising it. The work took three years and was physically testing and poorly funded. Her laboratory was described by one visiting scientist as 'a cross between a stable and a potato-cellar'; another noted with admiration that she 'worked like a man'. But her efforts paid off and at last she had isolated an element she named 'radium'; she had already discovered another element and named it 'polonium' after her native Poland. This newcomer was something even more exceptional. She explained that radium was three million times more radioactive than uranium.

Marie Curie's breakthrough sparked a frenzy of excitement that spread far beyond the scientific community. The historian of science Luis A. Campos described the initial impact: 'Unfathomably rare and intensely powerful, glowing in the dark and utterly unaffected by any outside force of nature as it gave off rays of unprecedented energy, radium was perhaps the most wonderful and perplexing thing the modern world had ever seen.'

By 1903, there were so many articles about radium that the secretary of the Royal Society noted that the 'newspapers have become radioactive'. At the American Museum of Natural History in New York, a sample of two grains of radium metal was placed on cotton in a glass case. It proved so popular that the police had to be called in to marshal the crowds. 'No other chemical element has ever attracted so much popular and scientific attention,' concluded Dr Lind some years later in *The Scientific Monthly*. 'Entire institutes have been devoted to its study. Medical clinics have been founded for its therapeutic use. Industrial companies have been formed and plants erected for its commercial production.'

A secret drawer had been opened in the universe's jewellery box, and in it was a supercharged gem. Everyone wanted a piece of it, to find ways to make use of its enhancements. Radium fertiliser made plants grow bigger, bloom brighter. Given to a group of flour worms, radium enabled one of the grubs to end up living three times longer than its life expectancy. 'It was as if,' wrote one clinician at the time, 'a human being should keep the appearance of youth for two or three hundred years.' Cosmetics companies developed radium beauty products. 'When scientists discovered radium,' one advertisement ran, 'they hardly dreamed they had unearthed a revolutionary beauty secret.' Crème Activa proclaimed that radium was 'a wonderful conquest of science in the service of beauty'. Hair tonic, face creams, anti-wrinkle treatments, complexion soaps all used the magic of radium to help people look better. Doramad Toothpaste was marketed in Germany with the tag line: 'Your teeth will shine with radioactive brilliance.'

In Paris, an 'afternoon radium cure' was introduced. It involved rheumatic patients taking afternoon tea in a room filled with radium vapours. Fashionable clients would gather to play bridge and they all reported a heightened sense of well-being. 'It is astonishing how many society women have discovered they are suffering from rheumatism,' reported the *New York Times*, 'in order not to miss the 3 to 5 o'clock "Radium Tea".'

Theatres staged performances where actors – decorated with radium paint – did 'radium dances'. At the New York Casino, the wonders of radium were presented in a skit called *Piff! Paff!! Pouf!!!* in which dancers performed daubed with luminous paint. In another corner of the same venue, 'radium roulette' was available, where the ivory pill bounced around a radium-painted wheel, sending out thrilling sparks of green before it settled. If you were down in the dumps, or tired, or flagging a little in the bedroom department, you could take a slug of Radithor ('CERTIFIED Radioactive Water Contains Radium and Mesothorium in Triple Distilled Water'), or a tonic of 'Perpetual Sunshine', or 'A Cure for the Living Dead'.

Radium spread its messianic light into the world's shadowy corners. A radium-based paint named 'Undark' was patented in the US and used first for military applications – night-sight compasses and flight instruments – and then for everything from fishing bait to light switches to glowing crucifixes. Wristwatches with luminous numbers and hour and minute hands proved madly popular. For millions of people, the glowing watch-hands were the attainable side of the radium craze. In the deep of the night, Americans raised their invisible wrists and the green shapes hovered in the darkness like clockface genies.

Medical uses of radium included the treatment of uterine fibroids, syphilitic ulcers, tuberculous lesions, warts, melanomas, gout, rheumatism, neuritis, ischias, diabetes, albuminuria, eczema, tumours of the lip and mouth, and 'every form of nervous complaint'. As many as 150 ailments were treated with radium. Alexander Graham Bell – he of the first telephone – suggested addressing internal cancers by brachytherapy, the insertion into the body of a tiny capsule of radium. Dr William Aikins, the first president of the American Radium Society, used radium to treat a wide range of conditions. One patient had a large tumour behind his left ear – 'a fungating mass covered with cauliflower excrescences'. After a few weeks of radium treatment, it had been reduced to a benign little ulcer.

Shortly after radium was discovered by Marie Curie and her husband Pierre, they lent a test tube of it to a friend, which he placed in his waistcoat pocket. It remained there for several weeks until he discovered that his skin beneath had become very red and inflamed. He told Pierre, and Pierre then tested it out on his own arm; the result was severe burning.

The perils of radium were not apparent at first. Or rather, they were if people had only stopped to look. But such was the wonder and so limitless the promise that the signs were largely ignored. A blind mania swept over the public, and George Bernard Shaw wasn't the only one to express surprise: 'The world has run raving mad on the subject of radium.'

If it was true that one of those flour grubs lived a very long time, it was also true that all the others died instantly. The radium pick-me-up Radithor worked so well that those who could afford to, drank it regularly. Eben Byers was an American iron tycoon, champion golfer and member of the East Coast social elite. He found that a quick drink of Radithor made him feel quite chipper. Soon he was drinking several bottles a day. He gave it to his lady friends and his racehorses. After three years, his bones began to disintegrate; his lower jaw collapsed. He died soon afterwards, and his body was so radioactive that when his lead coffin was exhumed decades later, the remains still brought out a crazed chatter from the Geiger counter.

A wider impact came with the 'radium girls'. From 1917, factories of the US Radium Company and the Radium Dial Company employed large numbers of young women to paint the dials of watches. Using camel-hair brushes, they were instructed to lick them – 'lip, dip, paint!' – to maintain a sharp point. Dozens became sick; many died. Their employers tried to blame it on syphilis and the girls' promiscuity, and those company directors fought vigorously against the lawsuits that piled up against them. The case that eventually found in the girls' favour helped establish US employee rights.

Some twenty years after discovering radium, with two Nobel Prizes under her belt, Marie Curie described the three years of intensive laboratory work as 'the best and happiest' of her life: 'I shall never be able to express the joy of the untroubled . . . atmosphere of research and the excitement of actual progress'. She was one of the heroes of early chemistry, embodying in her dedicated curiosity the heroic spirit of scientific enquiry. But during that time, she also recalled feeling exhausted, and was beset by bouts of illness, strange aches and torpor. In 1934, she died from aplastic pernicious anaemia, brought on by sustained radiation exposure. Even now, her notebooks, though decontaminated, are considered dangerous to work with.

Never have the hazards and benefits of the earth's matter been so intimately linked; never has a single element so brazenly offered life and death with the same hand. Radium is a substance from nature's top shelf, a forbidden fruit, a Pandora's box. It is the secret at the centre of it all, something so tempting, so powerful and dangerous, that simply to know of it is to sit with the gods. And that story never ends well.

One thing that redeemed the fad was the expense. Many so-called radium treatments contained no radium at all. To extract even the tiniest amount of radium requires huge effort. Several hundred tonnes of radium ore – pitchblende – had to be reduced in multiple ways, using vast amounts of energy, to yield about a gram of radium. After the First World War, that gram would cost upwards of $100,000 – more than three thousand times the price of gold.

In the early years, the only source of pitchblende was a mining town in the highlands of Bohemia. Joachimsthal, now known by the Czech name of Jáchymov, lies at the nexus of two major fault lines in a mountain range whose very name suggests mineral largesse. 'Erzgebirge' in German and 'Krušné Hory' in Czech both mean 'ore mountains', and their geology has had a striking impact on European and global history. In the Bronze Age, they produced tin. In the early sixteenth century, large amounts of silver were discovered, a horde of miners arrived and the town of Joachimsthal surged into being. The coins minted there were known as 'thalers', from the town's name, a word that through Dutch traders became 'dollar'.

To get to the silver, the miners had to shift tonnes of a blackish rock, which they chucked into the forest. That was pitchblende. And so it was that Marie Curie's discovery put Joachimsthal back in the spotlight, initially as a source of wonder, and then horror. From the waste of the Erzgebirge silver mines was born the atomic age – the thrill of radium, the miracle of nuclear power and the eternal shadow of nuclear weapons.

In the border town of Kraslice, not far from the ridge of the Erzgebirge, I exchanged a German train for a Czech train. In the shade of the station, a woman and a dog were dozing together on a bench. No-one else was about. I stepped over the tracks and into an empty carriage. Over the next hour a few other people drifted on, as if they were looking for a place to sit out of the sun.

Once going, the train picked up speed. Rail track and river weaved back and forth across the narrow Svatava valley, intertwining like Hermetic snakes. I stood at the open window and felt the breeze on my face, watching the slopes slip by in their springtime exuberance. We rattled over bridges, passed mossy wood-piles, dandelion meadows and grass-centred tracks. Ash and sycamore grew thick on the valley sides and everywhere was a green fluency – green pasture, green hedges and pale green leaves rising to the rim of sky. Sap was pressing up through each stalk and stem, flowing up bole and branch – while far below hot magmatic brines seeped through cracks in the broken rock.

Jáchymov is not what it was. Its population has shrunk. But it remains connected to its earthly gifts, or one in particular, which gives the old mining town a veneer of prosperity. Climbing down from the bus on the central roundabout, I could see three large white hotels: the Curie Complex Hotel (three hundred beds), the Behounek Spa Hotel (three hundred and fifty beds), and a little below them – built up against the valley side – the vast neo-classical Hotel Radium Palace (with its Madame Curie Suite, and a further three hundred beds). Built in 1912, the Radium Palace presented an ornate and elevated façade of some two hundred windows, the original temple to the ground's sacred powers. Each of the hotels depended on radon, the gas released by decaying radium, and each offered courses of radon baths, radon drinks and radon inhalations. That evening, every one of their nine hundred and fifty beds was occupied.

I found a room in a newish guest house and in the early evening walked up the valley and into the forest. I was soon high above the town's roofs, looking down on its elongated shape, a benign ribbon following the line of the valley. I remembered reading the hand-written memoirs of Helena Brochocka, a Polish noblewoman who came to these mountains to take the waters in the mid-1930s. She recalled the spa atmosphere and patrician ease: Italians and English, Germans and White Russians all mingling with the confidence of their shared class – no awareness of the carnage to come. There was also a veiled reference to an 'Austrian gentleman' who assaulted her on an afternoon walk in the forest.

Suddenly – CRA-CRACK, a double flash in the dusk, thunder bouncing around the hills. The first rain detonated in dust-puffs at my feet. By the time I reached the back of the Radium Palace, I was soaked. I pushed open the door, stamped my boots and shook my coat dry. No-one was around. I made my way through red-carpeted corridors, down a stairway, along another corridor.

There was music ahead: Bonnie Tyler's 'It's a Heartache'. In the darkened ballroom, a band was playing at one end, and a dozen or so couples were shuffling across the floor while others were sitting at tables and in banquette seats. There were chandeliers and columns, ornate stucco and gilding, walls hung with fruity swags of plaster, all multicoloured in the disco lights. At the bar, I collected a bottle of beer and found a table.

I couldn't help feeling that I'd stumbled on some sort of cult gathering, that these drinking and dancing couples had been inoculated against critical thought in order to share the same dangerous delusion. Like everyone who'd come of age in the 1970s and 1980s, I was conditioned to fear radiation, to file its calamitous effects under 'worst horrors'. I'd carried the projections of what nuclear apocalypse would mean from the pages of Raymond Briggs's *When the Wind Blows* and the made-to-shock films *The Day After* and *War Games*. I'd watched the night sky and wondered how would it look, that first flash on the horizon, and if you survived that, how it would be to live in a Mad Max world, to feel radiation pinging through your body.

Still, I'd been doing a little research, reading various articles about the effects of radon treatment. The results presented something of a mixed picture. In 2022, *International Journal of Public Health* published a piece entitled 'Cure or Carcinogen? A Framing Analysis of European Radon Spa Websites' (Geysmans et al.): 'Since the 1980s, radon has therefore been treated as a carcinogen, and is generally considered as one of the leading causes of lung cancer worldwide . . . it is remarkable that across the world there are facilities which offer services based on claimed health benefits of radon gas.' On the other hand, in *Rheumatology International*, the 2013 study 'Long-term Benefits of Radon Spa Therapy in Rheumatic Disease . . .' (Annegret and Thomas) concludes that for rheumatoid arthritis, there was a degree of analgesic benefit. A more recent article in *International Journal of Molecular Sciences*, 'Radon Exposure – Therapeutic Effects and Cancer Risk' (Maier et al.), offers a synthesis of all the peer-reviewed studies and articles: 'Radon therapy has beneficial effects on patients with painful, degenerative, and inflammatory diseases, describing a significant reduction of pain and enhanced mobility as well as increased quality of life.' It works, in other words, but so did Radithor, in the short term. The conclusion of Maier et al. was that the risks have not yet been fully explored: 'experimental research on the effects of radon exposure is needed on multiple levels'.

A couple sat down at my table. They had been dancing to 'Blue Suede Shoes'. 'Oj-oj!' sighed the woman. Sweat shone on their faces, and they glowed with good spirits. I raised my bottle and toasted them.

Hana and Marek had been coming for years. 'We visit in the spring, every year,' said Hana. 'Since his accident.'

Marek had been in a car crash in Slovakia. He'd injured his back and suffered chronic pain as a result. He duly reached for his shoulder, and winced.

'The doctor gave him tablets,' continued Hana, 'but the tablets made him ill. My sister told me about this place and, well, it worked at once.'

Marek nodded.

'Some people believe that radon's not safe,' I said.

'Oh, they are all qualified here. They know how much to give. Low-level treatment.' She gestured at the crowded room. 'Do you think all these people would come if it was bad?'

In the *International Journal of Molecular Sciences* article, the authors note that patients with chronic pain assess the risk of radon therapy differently to healthy individuals. Relief trumps the possibility of long-term harm.

'He always feels so much better,' Hana looked at Marek; a happy expression crossed his face.

They stood to dance again. I raised my drink and wished them good health. I watched them merge in among the other couples, all of them moving in unison to Abba's 'Dancing Queen', faces and bodies studded with the fractured light of a glitter ball.

*

At the top of Jáchymov is the Svornost mine, the town's cash cow. It started producing silver in 1518, loads of it. Vein after vein was uncovered, dipping ever deeper into the bedrock. Then it ran out. But arsenic, cobalt, lead and nickel took over, before uranium was found in the nineteenth century. It is one of the oldest working mines in the world – yet no rocks now come from it. It supplies the spa hotels with radon waters through long underground pipes. Approaching the mine from below, I could see the sheds and the headgear rising from the slope like some steampunk cathedral.

There I met Jiří. He was a large man with a large, open face. All his life he had worked in mines. Since two days after his fifteenth birthday, he had been going underground at various pits around the Czech Republic. He had come to Jáchymov and the Svornost mine in the late 1990s.

'Twenty-five years I have been here! I am seventy-one.'

'Really?'

I liked his bonhomie and his robust air of satisfaction, of a life well lived. We were standing in the mine's main building, peering in at some glass cabinets.

'Look!' He pointed at a grey tunic. 'Uniform. That was my apprentice uniform. I wore it about a year – then this one, black.'

In another case was a rock collection, and he took a Geiger counter and ran it in front of a lump of pitchblende – *tat . . . tat . . . tat-tat-tat . . .* 'Radioactive.' He was as proud of that as his old uniform.

Beyond the cases was a small cage. The shaft fell unseen beneath it. There was room for the two of us. Jiří hit a red button, a bell rang twice and the steel cabin wobbled and bounced. Then it started to drop.

'How deep, Jiří?'

'Five hundred metres.'

I pictured half a kilometre on the ground, upending it to have some sense of the space below our feet. Moisture glistened on the shaft wall. The cage was open-sided and it creaked as it crept downwards. Metallic clanging and a shivering of the cable, then a sudden high-pitched whirr, an acceleration and the rock was whizzing past. Jiří raised his hands to the bar above and looked down at his feet.

It took us several minutes to reach the bottom. We stepped out into a space where pipes and machinery spread out in pools of overhead light. Jiří pointed to the pumps.

'Pumps,' he said. He let his hand linger on them as he passed, as if they were living beings, workhorses. In the control room were a few screens, turn-of-the-century electronics. 'Computer.' He tapped at the keyboard to call up a schematic map of the mine. He pointed out the various levels and the location of the underground springs. Scrolling left, his finger pointed at the pipelines that led from the springs, through the mountain and down to three oblongs. 'Spa centres.'

We set off along the adit. The light grew dimmer. Our boot-scuffs echoed in the darkness. Water dripped from the roof and ran along a channel at our feet. Everything was wet. Somewhere above us, on a level 360 metres higher than this one, was a dam whose job it was to hold back the groundwater. A few inches of concrete was all there was to prevent catastrophic flooding of these tunnels.

Jiří turned to show me some metallic frames on the wall. They were covered in what looked like brown melted candle wax. 'Iron – very rusty.' After just five years, many of the structures need replacing. Radiation disrupts the atomic structure, accelerating rust – in one of those mysterious, life-like surprises of metals, some alloys are actually prevented from rusting by radiation.

'This tunnel is the same age as me,' he announced.

'You're in better condition.'

'No, I am old.' But in the half-darkness, I could see he was smiling. He ran one hand along the walls and let out a breathy 'Oo-oosh . . .' It was clear he loved this place, a love built from long familiarity, countless days spent down here as custodian of the forces that were channelled to the spa hotels, and drew tens of thousands of strangers each year to them.

We came to a side gallery – the Behounek spring. A wooden bathtub tied with steel bands stood to one side. 'This is the best water. I like only Behounek!' And he stooped to drink from a ladle.

'What about the bath?'

'It helps with pain.' He gave a stage wince. 'I have used it. Before they knew about the radon, the miners found that when they washed after a day's work, all their aching was gone. It was the miners who discovered how the waters worked!' He chuckled, and the chuckling ended abruptly, in a brief cough.

Down here, the radiation was at about two thousand becquerels per cubic metre, in summer about eight thousand. That was significant, but not dangerous. For me, and for the clients of the spa, it was fleeting. But for those who worked here, it was different; ten years of those levels would significantly raise the risk of lung cancer.

On the way up, we stopped off at a level close to the surface. 'Old-time mining!' There were traces here of the original silver workings. Everything was much smaller. We squeezed into a tunnel. I found my face close up against the rock.

'Think of working in here!' called Jiří.

The adit tapered and stopped – the end of the lode. The name 'pitchblende' is said to derive from *blende*– meaning 'deceive' in German. It looked ore-bearing but wasn't. The miners threw it away in frustration. But one era's waste is another's bounty, and the pitchblende later produced a new bonanza.

Up at the top again, daylight came as a relief. Jiří gestured to a row of high windows. 'Look – sun is shining.'

We both breathed in deeply.

'Fresh air!' grinned Jiří, and suddenly he was coughing again.

I walked back through the town. The main street was a wide boulevard of stepped buildings that followed the valley as it dropped towards the cluster of spa hotels. Up here in the older part, the buildings were opulent – proud façades from the eighteenth and nineteenth centuries with decorative mouldings and elaborate lintels and plasterwork. The colours were earthy shades of yellow and red, and together they signalled wealth, dividends paid out by the rocks on which they were built.

But only about every third or fourth house was occupied. The rest were in varying states of ruin. The stone was radioactive. A few were being pulled down and replaced – but the truth was that apart from the bustle of the spa hotels, Jáchymov was not a place people wanted to be. 'They think we glow,' said one resident. The population was a fraction of what it had been in the sixteenth century. A shadow hung over the town, like something shameful in its past.

In the early years of the radium craze, when Jáchymov was the only source of pitchblende ore, and the metal itself fetched over £1m an ounce, other sources were sought. In Cornwall, in the middle of the fishing town of St Ives, a redundant mine turned out to be full of pitchblende. The miners had always considered it low-grade copper ore and had spent years consigning it to the dumps. But with radium now in high demand, the tailings of Wheal Trenwith were scoured and the mine re-opened. A batch of St Ives pitchblende was sent from Cornwall to Marie Curie in Paris and she declared it of good quality. The Scottish chemist Sir William

Ramsay, who had recently won a Nobel Prize for work on the noble gases, claimed it was actually 'very much richer' than the Jáchymov pitchblende. He opened a radium factory in London and brought in crates of St Ives ore for processing. Once extracted, the radium was kept in a special safe of lead and steel with mercury-filled rods. Weighing a tonne and a half, the safe was only a metre high. It was stored in London's new Radium Institute, where Ramsay happened to be on the board.

At the same time, St Ives itself began to be spoken of as a possible site for a radium spa. The Radium Palace in Jáchymov was well known to be attracting hordes of well-heeled visitors from all over Europe. St Ives could do the same, adding sea air to the radium cures for a more complete sense of well-being. The branch line of the Great Western Railway would bring passengers to the stately Tregenna Castle Hotel – owned by the railway – which would serve as a centre for balneotherapy. Radium-rich waters would be piped from Wheal Trenwith to the hotel. Everyone would benefit.

The plans came to nothing. The First World War put them all on hold and by the time it was over, a little of the buzz had gone from radium. London's Radium Institute carried on into the late 1920s (it is now the Algerian Embassy), and Ramsay's radium factory was dissolved in 1921. In St Ives, the adits of Wheal Trenwith fell silent again.

A few months earlier, I'd had a call from my friend Sam the mine explorer. He lived in St Ives.

'The police,' he said, no prelims, 'are moving out of St Ives.'

'OK.'

'Selling up the station building for holiday lets. They're going up Camborne – for now, at least.'

'No more crime in St Ives, then?'

'It's the car park underneath. The mine entrance's right there underneath the bloody police station! I been waiting years to get in. You want to take a look?'

The ramp down to the underground car park curved as it dropped. Where police vehicles had once lined up in a rank of blue lights and luminescent livery was now empty space. Sam was already there, sweeping a Geiger counter over the concrete walls. There were a couple of others too – a man from the developers, someone from the town council and, in polished black shoes and collared shirt, the manager of St Ives's Royal Cinema, under which ran the mine's river – the Stennack (Cornish for 'place of tin'). Despite their presence, our mission still felt clandestine.

In the far corner of the subterranean space, a patch of sunlight came from the street above. An iron gate was silhouetted against it. Sam had already seen to the padlock with some bolt-cutters. Beyond the grille was the mine's rocky mouth. Dropping into it, Sam and I found ourselves in a large storm drain, a little less than head height.

'That's where the radiation will be – in the water.' He lowered the Geiger counter to calf height and the pings multiplied. '1.3, 1.4 . . . nothing too drastic.'

Half-crouching, we stumbled along the pipe. Our lights made pale rings ahead of us. The Stennack here was no more than a summer trickle, slopping at our ankles. The drain was new but it followed the old workings which had been dug for tin and copper in the early nineteenth century, then briefly for radium in the early twentieth. Where the pipe came to an end, deep beneath the town, the water was tumbling over a waist-high waterfall. I crawled up over it, into the rough-edged space of the mine. Sam passed me the machine.

'It's tight!' I wriggled on, keeping out of the water, then stretched out, pushing the Geiger counter up into the darkness.

143

'What's it say?'

'Not much above two.'

Over the sound of the water, I heard him scoff. 'There's more than that in town!'

All over St Ives the mine's waste had been used as a rubble base for tarmac. For forty years, the Stennack had run through pipes to provide water for much of the town. 'Didn't seem to do us any harm,' shrugged Sam. Maybe he's right. Studies in Cornwall show only a marginally increased incidence of lung cancer among smokers in radon areas, and there's also a slightly higher number of melanomas. I remember a TV news bulletin shortly after the Chernobyl nuclear disaster, reporting on the resulting radiation levels in the UK. A bar chart showed comparisons: London, Glasgow, Cumbria. At the bottom was a bar marked 'A Week's Holiday in Cornwall'. That bar stretched way out beyond any of them.

Later, Sam and I walked up through town to his house. It was a hot, cloudless day and the harbour buildings were bright against the blue beyond. We passed the Trenwith car park – known to all as 'the thousand-space car park'. It is built over the flattened-out dumps from the mine, and Sam's Geiger counter was louder there than it had been all morning.

Visitors were heading en masse to the town's beaches, burdened with deckchairs and bodyboards, children and inflatable animals. Sam let them pass, with a faint shake of his head. Behind them they left rows of glinting cars, the overspill of a town that was never built for cars. As well as the sun's rays, radium atoms were all around us, bouncing unseen into us, through us, spinning through mini-universes of cells and DNA, driven by an energy whose source is still as mysterious as life itself.

*

144

In the mid-sixteenth century, at the height of Joachimsthal's silver boom, the town's pastor was the charismatic Johannes Mathesius. His sermons drew thousands of miners to the newly built church – its steeple still dominant at the top of Jáchymov. Mathesius himself was a mineralogist and had made a good deal of money from mining. Often he would dress as a miner to deliver the sermons. Alongside the message of Protestant righteousness, he offered practical mining tips and broader ideas to help the miners understand the strange world they encountered below ground.

Mathesius's sermons were so popular that in 1562 they were bound and published in a book called *Sarepta* ('smelting'). In one of them he discussed metallogenesis – how metals 'mature' beneath the ground. All miners were familiar with the soft gooey substance known as *gur* that seeped out of the rock and crystallised in fissures. 'From this,' explained Mathesius, 'all ores and metals arise.' *Gur*, he told them, was the earth's seminal fluid, its seed spilling from certain places underground. It was an indication that, somewhere nearby, ores would be quietly maturing in their womb-like space.

Gur is now understood to be the result of chemical degradation of certain minerals – but at the time Mathesius's explanation fitted the widely held belief that the earth and its substances were in some way living. The language of geology retains the notion of life – rocks have 'veins' of ore, which are found in a 'matrix' (from the Latin for 'mother' or 'pregnant animal'). Overworked mines were often left fallow for the ore to regrow. The same view of the natural, in which all things were animate to some degree, was widespread in sixteenth-century Europe and acted as a guiding principle for that shady band of cosmological explorers, the alchemists.

The fathers of the Enlightenment dealt firmly with such whimsy. They separated the living from the non-living. Life came from life, in an organic cycle of sexual reproduction, and rocks were no part of it. If it was true that the new science had no explanation, other than the Book of Genesis, for where life originally came from, they did know this: the possibility of replicating nature's creative processes was ridiculous, the stuff of magic. Alchemy and all its gullible followers were pushed out into the cold.

Then came radium. On 20 June 1905, the front page of the *New York Times* read: 'Generation by Radium: Cambridge Professor Reported to Have Produced Artificial Life'. A few months earlier, in Cambridge's prestigious Cavendish Laboratory, J. Butler Burke had added 2.5 milligrams of radium bromide to some beef gelatin. Sterilised and sealed, the mixture displayed movement after a few days: 'a peculiar culture-like growth appeared on the surface, and gradually made its way downward, until after a fortnight . . . it had grown fully a centimetre'. Biologists in the lab flocked to examine the growth. They confirmed it was not bacteria, nor microbes nor yeasts, but something entirely unknown. Burke named it a 'radiobe'.

In the heady early days of the radium craze, Burke found himself a celebrity. Echoing Darwin's *On the Origin of Species*, he published a book about his work with radium called *The Origin of Life, Its Physical Basis and Definition*. For a short while the two books stood together in stature. In 1905, the *Daily Chronicle* called Burke 'the most talked of man of science in the United Kingdom'. But the more excitement people showed, the more sceptical the scientific establishment became. Sir Bertram Windle said that the radiobes were 'more like some aberrant process of crystallization than the behaviour of a living organism'. Sir William Ramsay dismissed Burke's ideas as 'mad'.

Many of Burke's actual claims were more tempered. Radium led him to a perception that harked back to a pre-modern age, back to animism. 'All matter is alive,' he wrote. 'That is my thesis.' James Lovelock's Gaia hypothesis has resurrected the notion, urging us to think of 'the biota and the rocks, the air, and the oceans as existing as a tightly coupled entity', its evolution 'a single process and not several separate processes studied in different buildings of universities'. Neither Burke nor Lovelock claims that rocks are alive as we are, but rather that 'life' offers a model for nature as a whole. Once a universal belief, the living earth – or Gaia – is something more than metaphor, but something less than literal.

In the early twentieth century, the discovery of radium and radioactive energy opened up a hitherto unknown dimension. It offered a thrilling set of treatments and tricks. Coinciding with the development of atomic theory, it helped to revolutionise ways of looking at matter, challenging long-accepted ideas. Atoms could decay and it was only their fast-moving state that made things appear solid. Some found the implications profoundly disturbing. Artist and critic Wassily Kandinsky recalled:

> In my soul the decay of the atom was the same as the decay of the whole world. Suddenly the sturdiest walls collapsed. Everything became uncertain, unsteady and soft. It would not have amazed me if a stone had melted into air and become invisible. Science seemed to me destroyed.

For scientists, the possibilities were so limitless that they often struggled to express themselves as scientists should. Frederick Soddy and Ernest Rutherford pioneered much of the early work on atomic decay. When he first saw thorium X turn into argon, Soddy recalled:

'I could feel my heart throbbing, and as though propelled by some outside force I heard myself utter unbelievable words: "Rutherford, this is transmutation!"'

To which Rutherford replied, 'Soddy, don't call it *transmutation*. They'll have our heads off as alchemists.'

Transmutation was the old fantasy of the alchemists, the idea that elements change into others, that base metal could be transformed to gold. Sir William Ramsay, who was involved with London's Radium Institute and was so dismissive of Burke's claims of life creation, had no qualms about using the same language:

> The transmutation of the elements no longer appears an idle dream. The philosopher's stone will have been discovered, and it is not beyond the bounds of possibility that it may lead to that other goal of the philosophers of the dark ages – the *elixir vitae*.

The discovery of radium re-introduced a universe of surprise, somewhere complex, infinite and mysterious. It also helped to focus attention on the micro. Dug from the earth, reduced through countless processes from its ore to a few pure and priceless grains, radium worked on the atomic scale – in the newly discovered world where particles were in constant motion. Looking through a spinthariscope – an optical device developed for examining radioactivity – one writer wrote of '"dead radium" displaying what suggests eternal life'. According to Luis A. Campos, the more serious work on radium was 'centrally involved in the study of the nature of heredity contained in chromosomes' and led indirectly to a new and more enduring 'secret of life' – DNA.

The radium fad left behind it the sulphurous whiff of quackery and a high body count – Marie Curie, Eben Byers, the radium girls. Soon it was put to even more deadly use. Four thousand years earlier, bronze changed the dynamic of warfare, and it didn't take long before radium's supernatural energy was applied to the field of conflict. With nuclear fission discovered in 1938 in Berlin, and

Jáchymov's uranium mines under Nazi occupation, it was feared that Germany would be the first to develop a nuclear bomb. Their programme failed. Those scientists who might have produced a bomb from Czech pitchblende could not sustain their commitment to the project. Some fled the regime, some were mobilised, while others were banished from the laboratory because they were Jewish.

Above Jáchymov, above the steeple of St Joachim's church, where Mathesius preached to the miners, and above the Svornost mine, I picked up a path into the trees. The spruce rose in perfect verticals. They were healthy here (there was no beetle) and armadillo cones lay scattered across the path. Squat shiny-leaf bilberries made up the herb storey, and in places the canopy was open where natural fallers had flopped against standing trees and their trunks cut cross-angles in the dappled light. No wind blew overhead. Everything was still and taut and warm. Unseen beneath my boots were a billion different things – liverworts and mosses and hornworts, lichens and wood-borers, fungi and bacteria, all feeding on each other, feeding off each other. A billion different things, or just one.

Out on top, I reached a high tarmac road. I stood and looked up the road, then down the road. In the distance I made out the faint stutter of an old tractor engine and, in due course, an old tractor. At the wheel sat a white-haired, white-bearded man. Standing on the plate beside him was a woman in a pale headscarf who gripped the back of his seat; in her other hand she balanced a wooden tray, covered with a muslin gauze. They came to a stop beside me.

'*Nikolaj tábor?*' I asked.

The man nodded. With his beard, he gestured to me to climb up on his other side. The back of the tractor held a box made of woven willow, like a large picnic hamper. We were barely going faster than I had been walking. Having no Czech, I tried German, then Russian. Conversation was not on offer. So we chugged on, the three of us, snail-pacing through the forest. After a mile or so, the man pulled the tractor to a halt, and nodded to the left. I offered my thanks and the woman pulled back the cloth to reveal a tray of red strawberries.

So it was with a fruity sweetness in my mouth that I stepped into the site of Nikolaj Camp. There wasn't much to see – a clearing long covered by grass, the creeping fringe of birch trees, but then – half-hidden in new growth – a wooden watch tower, a familiar shape from Second World War prison-camp movies.

First Hitler, then Stalin. Each of them craved a nuclear arsenal. The Nazis used captured Russians to dig Jáchymov's uranium; after 1945 the Nazis were the prisoners, forced to dig uranium for the Soviet Union. Soon the Czechoslovak government sent its own political prisoners to add to Jáchymov's convict workforce. Stalin then forced a deal, a treaty so secret that not even the relevant ministers saw it – and all the uranium from here was crated up and taken by sealed train to Russia. Jáchymov itself became a closed town. Throughout the 1950s, it remained a secret Soviet enclave.

Nikolaj Camp was one of a number of camps in these forests. Each one took their design from the Nazi Stalags. They had a double barbed-wire fence and a swept sandy perimeter, and barrack-like buildings for the inmates. Every morning the prisoners were tied five abreast and walked to the mines, a mode of transport they called the 'Russian bus'. They worked underground, inhaling radio-active aerosols, or in the processing plant known as the Red Tower of Death, where the breaking of the pitchblende meant even greater exposure. They returned to their forest camp by Russian bus. They died from the Jáchymov 'Miners' Disease', lung carcinomas.

Here in the forests above Jáchymov, all the early excitement of the atomic age – the crazier end of the health fad, the wild possibility of life creation – was finally snuffed out. I tried to square the benign-looking scene with the vanished areas of concrete and steel, to picture the solitary cells where prisoners, with professors and intellectuals among them, curled up in the impossible cold, the bunk-rooms and the bullying and humiliations, the nights filled with coughing, the malnutrition, the escape tunnel that a few of them managed to dig, and the commandant's neat and flourishing kitchen garden behind a high wooden fence.

Now there were several shades of vegetable growth and passerine song, and the stalks of a hundred waist-high poplar saplings. As I ambled back to the road, my toe tripped on something hard. I looked down to see a rust-brown nub. I kicked at the soil around it to expose a fitting or bracket. It was a piece of the old perimeter, the old iron caging apparatus – one metal employed to enable the extraction of another.

7

AEROLITE

THE FIRST IRON USED BY humans came not from the earth but the sky. Before the Iron Age and the development of high-temperature smelting, iron from meteorites was shaped for weapons and jewellery. When the invading Spanish asked the Aztecs where they got their shiny metal blades, they raised their arms and pointed to the heavens. The Sumerians called iron *an-bar*, 'fire from heaven', the Egyptians *ba-en-pet*, 'metal from heaven'. The word in Hebrew and Assyrian derives from *barzu-ili*, meaning 'metal of God'. A Hittite text identifies the terrestrial source of gold and copper, but explains that if you want iron, you must wait for it to fall from the sky.

One of the world's largest known meteorites fell on Greenland a couple of thousand years ago. The stone weighed some fifty-eight tonnes but broke into several fragments before impact. The Inuit learnt to cold-forge the iron for harpoons and flensing knives. In 1894, several lumps were tracked down by the explorer Robert E. Peary. He spent three years planning their removal. He built a short railway. He transported them to New York and his wife sold them to the American Museum of Natural History for $40,000 (about $1.3m today). Three of the meteorites are still on display, and the largest is so heavy that it is fixed by supports that go straight down through the floor, through the basement and the building's foundations, to the granite bedrock of Manhattan.

Meteorites were fairly scarce and chiselling them was not easy. Their iron was often rarer and more valued than gold, bearing in it the added mystique of its ethereal source. It was the stuff of princes and kings, linking their earthly bodies to the sky above. Nineteen objects of meteoric iron were found in the tomb of Tutankhamun – several iron daggers, a tiny iron headrest and an iron amulet; each was set in a frame of gold. In Egyptian temple inventories, iron objects were referred to as 'stars'. Himalayan singing bowls and *vajras* – ritual weapons – were made from aerolithic iron. The Black Stone of Mecca (*al-Hajar al-Aswad*) is set into the eastern corner of the Ka'aba – the very centre of the Muslim world. It was kissed by the Prophet and is now kissed by pilgrims as part of the *tawaf*. Although it is believed to be a meteorite, actual proof by archaeometric analysis has not been possible. The Omphalos stone at Delphi was likewise understood to have dropped from the sky.

At Emesa in Syria, a cult arose around a much larger black stone that had plunged to earth. In May 218 CE, its high priest was suddenly told that, following a coup, he was now Emperor of Rome. Marcus Aurelius Antoninus was his imperial name, though he retained his temple name of Elagabalus. The holy aerolite travelled with him to Rome, and the journey took the best part of a year because the rock needed to be regularly and solemnly worshipped. When they reached the capital, Elagabalus declared that from now on the stone would be the Roman Empire's chief deity. At summer solstice it was paraded in a chariot pulled by six white horses, and Elagabalus himself ran backwards before it in order to keep his eyes fixed on its glory. His rule was short-lived, undermined by his own liberal sexual appetites. After four years he was beheaded by the

Praetorian Guard. The sacred stone was returned to the temple in Syria, where it was smashed to pieces by Christians.

Common to almost every belief system – from Norse to Aztec, Phrygian to Berbero-Egyptian, Navajo to the Tengrism of Central Asia, and embodied in the metal heavens of the Finnish *Kalevala* and the bronze sky disc of the Únětice people – is the overarching significance of the sky. The heavens are a version of the earth, and vice versa: gold and copper underfoot, iron overhead – as above, so below. Not just materials but the same movements, the same forces and processes. Constantly in flux, the sky contains in it the eternal secrets of the earth. Each is a different side of the same coin, twin entities in a pattern that exists throughout the physical universe, where one thing mirrors another, and the genesis of one thing explains another, and the entire natural world works by a single set of laws which are constantly recurring, constantly repeating, varying and returning like the waves on the shore, like strains of melody, like the Music of the Spheres.

Look at the surface of the world and it presents diversity, infinite variation. But beyond it lies one big shining unity. Invisible in normal circumstances, it is a truth confirmed in moments of transcendence or ecstasy, or in finding a weighty pebble of iron and nickel and realising it's just a chip from the heavenly ceiling.

As above, so below – an idea distilled in antiquity and carved into a boulder-sized jewel of a stone. The Emerald Tablet was revered in the Arab world and in Europe alike for its green translucence and the words carved into its surface. As an object, it never actually existed, but rumours of it were powerful enough, as were the ten maxims it displayed. They were the teachings of Hermes Trismegistus, Egyptian priest-king. He never existed either – but during the Renaissance it was believed he had been given the primal wisdom long before the Pharaohs, before religions diverged and the pure single truth was forgotten.

The Emerald Tablet begins with his assertion of heaven and earth's essential likeness, and among those who translated it was a secret admirer of the speculations of Hermes Trismegistus – Isaac Newton. His version begins:

Tis true without lying, certain and most true.
That which is below is like that which is above and that which is above is like that which is below.

The rest of the tablet's text is fairly abstruse but its slippery language suggests theories of universal concord, celestial movement and the possibility of the transmutation of matter. For hundreds of years these ideas occupied the probing minds of Arab, Jewish and Christian philosophers. Similar beliefs cropped up in early China and India. Terms used were many – the *Magnum Opus, Rasayana, Waidan* and *Neidan,* chymistry, the Hermetic tradition and *al-kimiya* ('alchemy') – or, the search for the Philosopher's Stone.

Pursuit of the Philosopher's Stone was a lifetime's occupation, years of sustained wondering combined with methodical enquiry. It stemmed from the conviction that the world around us is a place of pattern and hidden powers, and it's all moving towards perfection – the soul of humans no less than the stuff of metals. No nobler activity existed for a thinking layman than to try and peer into the mysteries and discern the one beating truth from which all things come, and the first place to look was the sky. The sun and moon and five known planets corresponded with the seven known metals. By studying the movements of those bodies in the sky and manipulating those substances on earth, the *prima materia*, the Philosopher's Stone, would ultimately be revealed. As above, so below.

'Every single thing that we value on earth,' says internet billionaire Naveen Jain, 'is in abundance in space.' Jain, who was brought up in Uttar Pradesh and made a fortune in Silicon Valley, has built up one of the world's largest collection of meteorites. His company Moon Express also acquired the right to mine metals on the moon, whose surface is pitted and scarred from the impact of millions of tonnes of precious minerals. The solar system is full of lumps of metal, spinning and hurtling through the emptiness. One, the size of Manhattan, orbits continuously between Mars and Jupiter. If it struck the earth its metals would be worth enough for everyone to receive a personal dividend of $1.3bn. Except we'd all be dead.

Rocks from outer space have, at various times, destroyed life on earth. But they may also have brought it. Meteorites have been shown to contain complex molecules – amino acids, extending possibly to complete proteins. A Japanese expedition to the asteroid Ryugu returned with samples containing uracil (a building block for RNA) and niacin (a vitamin essential for life). Such substances, suggested the project's lead astrochemist, 'may have been provided to the early Earth as a component of asteroids and meteorites [and] had a role in prebiotic evolution on Earth and possibly for the emergence of first life.'

In the medieval kingdom of Bohemia, not far south of Jáchymov, a meteorite fell to earth. It was at once identified by the local people as an incarnation of the evil burgrave who had once tyrannised them. They chained the stone up in a dungeon, lest he escape and make their lives miserable again. Another story from Bohemia tells of a sky-fight between a rebel angel and the Archangel Gabriel. In the story the rebel angel receives a hefty punch from Gabriel and crashes to earth. The impact made a large dent in the ground. As the rebel fell, a chip pinged from his jewelled crown – a single piece of vitreous rock flashing through the heavens. It embedded itself in the soil, and on that spot – in Prague, in the district of Hrad – was built the imperial castle, the largest ancient castle in the world. In its courts and galleries, from late in the sixteenth century, the Habsburg ruler Rudolf II divided his time three ways – ruling the Holy Roman Empire, contemplating the beauty of his art collection and discussing esoterica with his alchemists.

In 1989, two professors from Boston University examined a set of satellite images taken from 22,500 miles above the earth. They identified an impact site in the west of what was then Czechoslovakia. It measured some two hundred miles across, and roughly in its centre lies the city of Prague.

Down to Prague on a hot afternoon, on a hot bus, down from Jáchymov and the wooded slopes of the ore mountains, across the baked plain to the Vltava river and a guest house near the city centre. I climb to my top-floor room, push open the window and sit there, leaning on the sill, in one of those slots of time when you want to be nowhere else. The city from up here is all orange roofs and angles, a thousand different ways to underpin the sky. The Church of Our Lady before Týn rises high above it all, twin spires pushing for something and never quite reaching it, masonry black with age.

In the early evening, I crossed to the left bank, then looped up past the castle and back over the river. The city was busy. Heat filled the street-side bars with visitors and pulsed from every stone – from the Carboniferous arenites of Charles Bridge, the *opuka* limestone of the Hrad district and the quartzite cobbles of the Old Town. Prague's relationship with rocks goes back centuries, running through its story like a streak of crystal in grey shale. Unlike the story of Jáchymov, and its minted silver and nuclear bombs, Prague's tale is one of gems, aesthetics and the complicated search for the *prima materia*.

I had come to Prague not to look for the Philosopher's Stone itself, but to try and understand the world view of those who did. I had the names of several people to chase up – scholars and writers, friends of friends. Searching for anyone to talk to on the subject proved as elusive as the stone itself. 'No time,' said one medieval historian, 'I am like mirage.' 'To me is given many duties,' explained a professor. 'Not all pleasant. I cannot meet.' One simply texted: 'Covid.'

With a couple of irons still in the fire, I took to physical exploring and specific reading. I visited the *Sál minerálů* (mineral hall) of the National Museum and marvelled at its sweet-counter display of four thousand rocks, and the darkened room dedicated to its meteorite collection. I toured the castle and its grounds. I had some exhilarating sessions in the faculty libraries of Charles University. A little of the city's surface dropped away.

Step back a moment, away from the now, from the heat and summer crowds, the ringing trams and electric scooters, the students lounging in the shade of small parks, the beer-drinkers and yoga-stretchers, the shepherded flocks of visitors, the weary office workers waiting at pedestrian lights. And picture this – the city in the late sixteenth and early seventeenth century, one of Europe's great capitals, riven by confessional disputes like everywhere else, but in thrall too to the shadowy realms of cosmic speculation.

The Bohemian city of Prague was then the centre of the Holy Roman Empire. The land around it was rich in minerals and metals but also in the sort of colourful crystals and rocks that pleased the eye and ignited the imagination. This was a country, goes the proverb, where the stone you use to hit the cow is worth more than the cow itself. Bohemia's stones held in them a kind of magic, a celestial power and mystery that contributed to making Prague at the time a hub of all things half-known and wondered at.

Step back, because to see the world as it was seen then requires relinquishing the world as we see it now. It needs an open mind to accept on their own terms things like natural magic, transmutation and alchemy, to consider them not as some risible delusion, a cul-de-sac of counter-empiricism, but instead a view of the natural world that can offer us a chance to re-examine our own, to shift our glance from received wisdoms and see it all afresh.

The search for the Philosopher's Stone always attracted its fair share of cranks and crooks. From the time it was first mentioned in medieval Europe, the elusive substance was tainted with tales of corruption and greed and vanity. But there were those who were more diligent, more idealistic, who dedicated their lives to a pursuit whose professed object they knew was probably unattainable. It was a quixotic quest, but also an edifying use of the mortal span they'd been given, a life of enquiry and experiment. Then came the Scientific Revolution, and from the late seventeenth century onwards, alchemy and the Hermetic tradition were firmly skewered.

Yet sift through the broken alembics and cloudy beakers, the scattered manuscript pages with their arcane symbols and slightly bonkers formulae, and there are precious notions in there. They act as correctives to the wild imbalances of Enlightenment dualisms – mind and body, animal and vegetable, mineral and organic, living and non-living. We have learnt to value taxonomy: the naming of plants and beasts, rocks and stars; the distinguishing of one thing from another. In doing so we push to the margins a profound truth, one recognised by Goethe and others who brushed with Hermetic thought, that the world is characterised less by its parts than by the connections between them.

Alchemy in sixteenth-century Europe, and in Rudolf's Prague where it thrived, was a part of the more general field of 'natural magic'. Magic – like alchemy – is not a word to use now in polite academic society. Yet at the time, the term simply applied to all that was inexplicable in the material world. The twenty-book *Magiae Naturalis* by the Neapolitan Giambattista della Porta beautifully sets out its principal questions. In them – in nascent form – are the concerns that still occupy the disciplines of natural science:

What are the effects of fire, Earth, air, and water, the principal matter of the heavens; and what is the cause of the flowing of the sea, and of the diverse coloured rainbow; and of the loud thunder, and of comets, and fiery lights that appear by night, and of Earthquakes; and what are the beginnings of Gold and of Iron; and what is the whole force of hidden nature.

Magiae Naturalis was a hugely popular book in early modern Europe. It was translated widely. A contemporary enthusiast was Rudolf II, who corresponded with della Porta. The emperor was never much good as a ruler but as an active advocate for natural magic and Hermetic ideas, for celebrating and exploring all that was strange and wonderful in the world, his commitment was exemplary.

One afternoon, I crossed the river and climbed to Prague Castle and the gallery that once housed Rudolf's art collection, his *Kunstkammer*. Among the rump assortment of paintings, I fell into conversation with a retired teacher from Brno. 'These are nothing,' she scoffed. 'Once we had the greatest collection in Europe.' She spoke of Prague in the time of Rudolf II as 'the heart of the world' and was admiring when she described Rudolf himself as 'like an ugly god'.

King of Bohemia, King of Hungary and Croatia, Archduke of Austria, Holy Roman Emperor, Rudolf was never comfortable with politics, the whispering and feuds, the factions and divisions. Whenever he could, he came here to his *Kunstkammer* to gaze at his collection – Titian and Spranger, Heintz the Elder, Breughel the Elder, Bosch, von Aachen. Dürer's *Feast of the Rosary* was carried for him over the Alps, upright to avoid damage, by four men. The 'world's greatest patron of the arts', Rudolf was known sometimes to sit motionless before his beloved pictures for hours on end.

The *Kunstkammer* was not merely an art collection. It was also a *Wunderkammer* – a gathering of natural wonders where Rudolf had created a microcosm, a condensed version of the universe in all its miraculous beauty, natural magic in collectable form. In countless cabinets, in chests of walnut and ebony and ivory, were the weird glories of nature: hippo teeth, bezoar stones, dissected frogs, elaborate antlers and fossilised sea creatures, Seychelles nuts made into ewers, a unicorn's horn, the foetus of a dragon, the jawbone of a Siren, a pelt of fur that dropped from the sky.

A very large proportion of the content of Rudolf's *Wunderkammer* was stone. He was devoted to glyptic and lapidary art. From the beginning of his reign, he had been dispatching agents around Bohemia – to the mines, to the mountains, anywhere the glittering materials of the earth might be found. The success of the agents was evident all around him. Jasper, chalcedony, Bohemian garnets, bloodstone – all brought to Rudolf in his *Kunstkammer*. Anselmus de Boodt was one of those agents and also the royal physician, and he explained his master's belief in the devotional significance of rock-collecting:

The Emperor is a lover of stones, and not simply because he hopes thus to increase his dignity and majesty, but through them to raise awareness of the glory of God, the ineffable might of Him who concentrates the beauty of the whole world into such small bodies and in them unites the seeds of all other things in creation.

Each stone in the microcosm of Rudolf's *Wunderkammer* was a microcosm too. In every crystal, every formation of oxidised metal or sulphate, was the essence of everything else. In thirty-seven cabinets, stretching for sixty metres along the walls of the *Kunstkammer*, and in the hundreds of drawers and compartments were his *Naturalia* – his minerals and gems – and all of them possessed in them some latent power. Contemporary thinking considered them the basic materials from which all things living and non-living come, in the same state of flux as everything else, involved in a slow process of transformation that could be accelerated by chemistry or by craft.

Rudolf's greatest commission, his greatest transformation, was a table – a square table made from coloured stones. The stones had been gathered here in Bohemia and sent to Italy. There, over a period of six and a half years, they were cut and finely fashioned into a smooth surface. The table was less a piece of furniture than a version of the world itself. 'The whole cosmos is present,' wrote the art historian Karla Langedijk. De Boodt at the time described its effects: 'Mountains, woods, trees, cities, rivers and clouds . . . were represented "as though painted" in precious stones inlaid invisibly.' He called the table the 'eighth wonder of the world', adding that 'it can without injury be compared to the Temple of Diana at Ephesus'.

When Rudolf stood to leave his *Kunstkammer*, his favoured route led not to the Great Hall with its ministers, petitioners and ambassadors, but down the corridors, along inside the northern ramparts, within earshot of his birds, the 'finest aviary in Europe' with gyrfalcons, pelicans, flamingos and breeding dodos, and beside it the Lion's Court with its lion cub presented by the Ottoman sultan and much beloved by Rudolf. There were also tigers, bears, leopards and cheetahs (with their always female keepers) – until he reached the Powder Tower. Here ores and metals from Bohemia's mines were stored in barrels, with powders in leather sacks. There were kilns and glassware, trivets and tongs, and shiny-faced men working in the heat. The Great Work was underway, the constant crushing and blending, boiling and distilling, the tireless quest to break through the earth's mixed-up materials to discover the Philosopher's Stone.

In the grounds beyond the castle, experimental forges and furnaces were built, for lack of space, in the flower beds. The city beyond housed any number of sites of enquiry – smoky basements, rooftop observatories, makeshift libraries. Some were secret, some of royal commission. Some practitioners were true believers, some silver-tongued chancers. The great Giordano Bruno was in Prague for a while, publishing a book of esoteric lore and dedicating it to Rudolf, preaching his own version of the Hermetic tradition – the primacy of the human imagination, the cyclical nature of change and the revolving of the earth around the sun (he would be later burnt at the stake in Rome). The Polish alchemist Michael Sendivogius, who helped design the emperor's mines, was said to have performed for him a lead-into-gold transmutation. Astrologer to Queen Elizabeth I, John Dee was here with the crystal given to him by the angel Uriel, and along with Dee was his shady accomplice Edward Kelley, who had no ears but was in possession of a red powder from a bishop's grave at Glastonbury Abbey which contained traces of the Philosopher's Stone.

At the same time, Rudolf's official mathematicians, Johannes Kepler and Tycho Brahe, were conducting groundbreaking work on the movements of the planets that helped lay the foundations of astronomy as a science. They too were not at all averse to Hermetic principles nor to practising as astrologers. Tycho Brahe foretold the death of Rudolf's lion and that it would coincide with the emperor's own demise.

Prague under Rudolf was a halfway house between medieval and modern Europe. Up to two hundred alchemists were practising in his capital. It was probably the last moment when astrology and astronomy, numerology and mathematics, herbalism and botany, alchemy and mineralogy, natural magic and empiricism could all sit down together at the same table.

In August 2002, a once-in-five-hundred-years deluge struck Bohemia and swelled the Vltava river to such a height that its waters spilled over into the streets of Prague. Among the many damaged sites was a home appliance store in the Old City. As the levels dropped and the clean-up began, the store's basement wall collapsed and exposed a series of unknown cellar rooms. In them were preserved ancient stone furnaces and glassware and manuscript pages covered with formulae – and all of them intact, as if a spell had been cast to remove the rooms from time's ledger. The rooms were at once identified as an alchemist's laboratory from the time of Rudolf II. The shop was sold to an entrepreneur, who ditched the home appliances for something much more lucrative – the *Speculum Alchemiae* or Museum of Alchemy.

A woman in a starry top was on duty and when I asked to see the museum, she locked the street door behind me. I followed her through to a back room, where a bearskin rug covered the floor, a wheel chandelier hung from the ceiling and a stuffed alligator lay on the mantelpiece. Period mystery.

'This house is one corner of the magical triangle of Prague,' explained the woman in her well-rehearsed spiel. 'Vyšehrad – the old castle – is one point. The newer castle, where the great Rudolf ruled, is the second – and the third is exactly here, where you are standing . . . A legend says you can sometimes see a flaming carriage dash through the square outside, pulled by goats.'

I nodded politely.

'And now . . .' She reached into a bookcase, twisted a rampant dragon, and the bookcase swung open. 'The alchemist's workshop.' Stairs led down to the vaulted cellars, and in the catacomb light were boulder-built ovens and bellows and dried herbs, flagons and barrels and shelves of elaborately shaped flasks – Erlenmeyers, Florences, Pelicans. Upstairs in the museum shop it was all kitsch – Elixirs of Love and Memory and *Aurum Potabile*, made from recipes in the manuscript pages, from 'herbs gathered in monastic gardens on moonlit nights'. The rooms below felt a little more authentic.

But in truth, alchemy in Prague is now nothing more than a visitor attraction. Alchemy-themed bars, ghost tours, costumed re-enactment – a peeping-in on something still coloured with taboo and intrigue. Alchemy was always prey to popular prejudice, its nobler principles hidden behind rumour and caricature.

If Rudolf's search for the Philosopher's Stone has left a legacy of kitsch, his enthusiasm for actual stones has more genuine heirs. I sought out Prague's gem stores. The first was in an art deco building but it was boarded up. The second boasted 'Bohemia's finest collection of minerals and gems' but was now an empty window. The third proved more fruitful – a shopfront full of malachite and lapis and amethyst. I pushed open the door and began to browse boxes of polished stones and a floor-standing display sample of a geode.

The shop belonged to Marta. She was sitting behind the counter, sunglasses pushed up high on her head. Most of her trade, she explained, was in healing crystals. She pulled out drawers and drawers of them. Trained as a geologist, Marta was somewhat sceptical of the claims but not averse to helping satisfy them. She had worked for a while at the Czech Geological Survey but the technical work bored her. Last summer she chucked it in, took out a loan on the store and filled it with exotic stones. In its colours and shapes, I felt the pull of my old obsession.

Marta was speaking of her stock, her own adventures rock-hunting and the country's subculture of rock collectors. 'Those who like minerals here, they live for them, they love them. Sometimes—' she inverted a finger to point at herself, 'maybe, a little too much.'

It was Marta who first showed me a gem that has a very unusual provenance. She unlocked a cabinet and picked out a small green stone. She held its anvil shape to the light. 'This stone is very special.' The sun from the street highlighted its bobbled surface and its deep vitreous green.

'Is it a beryl?'

'No – it's silica-based, like glass. It's a moldavite – formed by a meteorite. Only here in Czechia, in one small area, can you find them.'

'Where's that?'

She took out her phone and on Google Maps zoomed in to the south of the country, where the Vltava river was much narrower, near the town of Český Krumlov. 'Around here.' Her long fingernail, as she tapped on the place, was moldavite-green.

The next day I crossed the river and visited the Czech Geological Survey. A dusty lump of petrified wood stood in the entrance hall. In the library – a small, tucked-away room – was no-one but the librarian. I asked her what she had on moldavites.

Moldavites are among the world's rarest minerals. They belong to a group of rocks known as splash-form tektites. They are not strictly aerolites yet are formed by them. In the split second before a large meteorite strikes the earth, the heat and pressure are so intense that material on the ground is turned instantly to ash or gas or liquid. Any silica melts and is thrust upwards in sudden drops of fluid glass that carve an arc high into the atmosphere and beyond, peaking in space before gravity takes hold and the droplets fall and harden, vitrifying into tear-like shapes as they re-enter the atmosphere.

There are fewer than two hundred known impact craters on earth. Out of these only about four have produced tektites, and of those four all but one are tektites of an obsidian-like black. For reasons to do with nickel content, the tektites that fell primarily on a couple of spots in southern Bohemia were not black but a beautiful translucent green, and these are moldavites, named after the German 'Moldau' – the river that rises near the German border, flows north through Český Krumlov and Prague as the Vltava, and on into the Elbe.

The meteorite that resulted in moldavites was nearly a mile across. It fell fifteen million years ago in what is now Bavaria. Striking the ground from the south-west at an angle of about 55°, it sent the debris splashing north and east. The green silica drops landed hundreds of miles away. Over time they were washed down into streams and sediment basins and then slowly encased in clay. Moldavites are affected by the acids in subsurface waters, so about 99% of them have been lost. Those that are left often reveal the chemical erosion with a wonderfully textured surface, an intricate network of sharp ridges and gullies.

The longer I read about moldavites that morning, and the more I thought about them over the coming days, the more they bounced around in my head in a kind of associative fascination, a Brownian bundle of recent notions and images. There were aerolites and gems, meteorites and *prima materia*, all tangling with the emerald-green tablet of Hermes Trismegistus and its inscribed truths, with Archangel Gabriel's space-fight and the rebel angel whose crown shed a jewel-shard to fall to the ground beside the Vltava river, and with the Philosopher's Stone and its echoes of celestial process on earth and earthly process overhead. And I kept picturing this:

the moment a lump of sky hit the ground, as a bullet strikes mud, a gigantic, superheated splash – and then, many miles away and minutes later, a shower of glassy green stones.

Show me a finer example of natural magic, and I will be duly amazed.

The following morning was wet, and I took an early train from Prague, heading south through dripping forests, past mist-wrapped hilltops and damp smallholdings neat with bean-rows. The heat dome of the past weeks had cracked. It was now all but replaced with cloud. I was heading for Český Krumlov, moldavite country, but on the way, in the old mining town of Příbram, there was a mineral fair to be visited.

The town came into view with cottage outskirts, allotments, then stands of Soviet-style apartment blocks. On the edge of Příbram was Svatá Hora – 'the holy mountain' – on which stands a church featuring a tiny miraculous statue of the Virgin. The statue was set in an altar of sheet silver. Příbram had been part of Europe's sixteenth-century silver boom and was given royal status by Rudolf II, and its mines poured money into the state's accounts for centuries. Then they all closed, suddenly, in 1989, with the Velvet Revolution.

In the town's *kulturní dům* ('culture hall'), at the end of the main hall was a stage and above the stage a small cartouche of a miner's lamp and a hammer drill. The mining was gone but it left in its wake a fascination for rare rocks. The hall and adjacent rooms were given over to rows and rows of trestle tables. Each one gleamed with a mineral display. Crowds shuffled down the aisles – family groups looking for beads and brooches and palliative crystals, several generations with varying levels of interest in the stones.

Mainly though it was men – men with indoor complexions and close-focus stares, bookish and suited men, men with ponytails, hunched over eyeglasses, bow-legged men in singlets or portly men with T-shirts pulled tight over round bellies – *Pink Floyd, Guns N' Roses, Staropramen Pilsner*. Behind the tables were versions of these same men, the pedlars, with their spreads of cherished rock.

The range was spectacular – galactic in scope, geological in timescale: blue aquamarines from Pakistan, opals from Australia, purpurites from Namibia, Canadian trilobite fossils, yellow citrine crystals from Brazil, orange cerussite from Democratic Republic of the Congo, purple amethyst from Zambia, black tektites from Vietnam. Several stalls were selling bits of aerolite – including fragments from Argentina's Campo del Cielo meteorite, and slices of the Sericho pallasite meteorite from Kenya – whose appeal was partly its celestial source and partly its resemblance to nut-cluster nougat.

I wandered up to the balcony for a bird's-eye view. The crowd had been building through the morning and there was little visible of the polished floor below. The hall was not overlit, but scattered through it were pools of halogen light and the technicolour of the minerals. It put me in mind of a flower stall or spice market – but these commodities had never lived. They are the most striking things the non-organic earth has to offer. Vast landscapes, sculpted cliffs, rounded boulders – none of them can quite compare with the intense vibrancy of such stones.

The first evidence of rock-collecting comes from a few million years ago: a piece of jasper gathered by an australopithecine in southern Africa. Ochre was also sought, and later gold and lapis, cobalt and copper. As the Enlightenment gathered pace in Europe, stones became objects of desire for grandees and aesthetes, a version of the medieval fascination with possessing religious relics. Lady Clive and Georgiana, Duchess of Devonshire left substantial mineral collections, as did John Ruskin. Catherine the Great had her Amber Room and her Agate Room and her Brilliant Room for jewels. In the US, banker J. P. Morgan was a keen rock collector and had a pink beryl named after him. Archduke Stephan von Habsburg-Lothringen was 'one of the greatest collectors of all time' and a silver antimony from here in Příbram was named stephanite in his honour. Tsar Alexander I's collecting gave rise to alexandrite, Goethe's to goethite.

All of these collectors were attracted by the unusual, the brightly coloured. Such collecting often became fashion, the competitive quest to outshine others. But at its core the appeal was something less prosaic, a curiosity about natural process. What hidden force created these stones? What goes on inside the earth, or in the sky, to produce such things? Rocks are a portal to the world behind this one. Goethe realised that after his first visit to the Harz Mountains. Marie Curie did too, as she held a lump of pitchblende and realised there was something dramatic going on inside it.

As a child, I remember the thrill of discovering a lump of houndstooth calcite in an old quarry, an amethyst in a laurel bush. Examining the impulse then had no relevance. Now I was aware, looking down on the scene below, of a negative reaction. What prudish disapproval was that? Was it the commodification? Or the profligate ease of their far-off provenance, wonders gathered from every corner of the globe and so casually exchanged?

Behind one stall, I met a man who didn't quite fit the mould. He wore a straw hat and denim shorts. He looked like an actor – not a particular actor, but a generic actor. An eyeglass hung from a chain around his neck. Speaking softly, he itemised the rocks before him: 'Kinzigite ... pirnelite ... bludovite ... arsenopyrite ...'

His stall lacked the colour of the others' – but it had something theirs did not. Now I understood my disapproval of the other tables.

These rocks he had actually found. The appeal for him, and for me, was always more in the finding than the keeping.

'I go into the hills near my home, and at other sites in the country. On other stalls—' he gestured around the hall, 'they are mainly business.'

He was a gold prospector by profession, working mainly in West Africa. 'Benin at this moment. I am lucky,' he gave a film-star smile, 'because my hobby is also my job.'

I asked if he had any moldavites.

He scratched his chin. 'Moldavites are a problem. You will see many for sale here but—' he glanced to one side and dropped his voice, 'they are mainly fake.'

A few hours south, and the old town of Český Krumlov turned out to be something of a gem itself. I entered on foot, beneath a short viaduct strung between two cliffs – between the high castle on one side and its gardens and outer buildings on the other. The whole structure was spectacular – a covered bridge three storeys high, itself supported on several layers of arches.

The town beyond lay across a simpler wooden bridge and was a cluster of medieval buildings bunched inside a looping bend of the river. In one of its cobbled streets was the *Muzeum vltavínů* – the Moldavite Museum. There I'd arranged to see Vít, who had set up the museum. I found him sitting outside, next to a pram.

'It's alright,' he said. 'She's asleep.'

I bent to look in on a swaddled little face, with the pearly rightness of babies' skin. 'She's beautiful.'

Vít himself looked tired, young-parent tired. It had been several years since he'd left Prague for Český Krumlov. As a boy, he explained, he'd loved two things – stones and photography.

At school he'd helped catalogue and display the mineral collection. He went on to study geology at Charles University in Prague. Soon after first seeing moldavites at a mineral fair, he realised university was not for him. He chucked in his degree and joined an advertising company – but moldavites were always at the back of his mind.

'It was never about owning them. Or their power. I just liked their story.' So now he found himself here, married to a local nurse, with a baby, running a moldavite museum. 'Strange how everything happens,' he said, with a half-smile.

I went inside. The stones themselves were laid out in sand, in glass cases. Their presence was partly the celebrity aura of them, their rarity, and partly their innate form. Arrowheads, sausages, medallions, each one conceived in the physics of liquid in space, moulded and hardened in the trajectory, shaped in the fall to earth, shaped again by the action of acidic water. Some had a surface of spiky relief and these were known as 'hedgehogs'.

In the very few villages where they're found, moldavites have long been ploughed up in the fields. One or two have appeared in prehistoric sites – notably in company with the Upper Palaeolithic Venus of Willendorf in Austria. Until recently, no-one understood their global scarcity nor the celestial role in their formation – but they did find them appealing. They were used as pendants, inlaid on walking-stick tops, given as gifts to hopeful brides. Then in the late eighteenth century, the first study of them hinted at a connection with meteorites. A hundred years later, as Europe stood on the threshold of mass consumption, moldavites were displayed at the 1891 Prague Jubilee Exhibition, and for a short time they became a gem of choice in Paris and London. Prices rose, and dealers began to turn up in Bohemia's southern villages asking for the stones. Demand soon outstripped supply – until it was discovered that green tektites could be very convincingly created by using bottle glass. They fell from favour.

Now in the twenty-first century, an intuitive belief closer to alchemy lies behind a revival in moldavites. Advocates of crystal healing hold the stone in high esteem, and dozens of websites explain the cause of its benefits. According to one: 'A powerful symbol of our connection to the cosmos. It is thought to offer a direct link to the energies of the Universe and to help facilitate major spiritual growth and transformation.' Moldavites' effects are reported to be palpable: 'It is not uncommon to see people faint or feel light headed due to its intense metaphysical energies.' In an eight-month period in 2021, *#moldavite* clocked up 280 million views on TikTok. Renewed popularity has brought a fresh wave of fakes.

Wearing or carrying a moldavite is claimed to re-shape your life, helping to break down obstacles, precipitate necessary change, eliminate the ephemeral, purge the peripheral. Empty relationships are thrown aside, untaken journeys taken, untold truths told, worthless jobs left, start-up businesses started up. Procrastinations vanish and many put-off pregnancies are put on. Users should also beware: 'If you're in a bad place, now is not the time to grab this tumultuous tektite.'

Like all the earth's sought-after materials – like oil, gold and peat, radium and silver – the digging of moldavites has had mixed consequences. Extraction brings value to consumers and money to producers, while those who live at the source see their land ripped open and plundered. The forests in southern Bohemia soon became crowded with moldavite hunters, the ground peppered with pits. The authorities stepped in, pointing out that it was illegal to dig for moldavites. For a while, police with dogs patrolled at night, but the trade resumed as soon as they slacked off. Local people were advised not to approach the diggers. Hunting for the green jewels, they were told, had made these people aggressive.

Vít was pushing the pram back and forth when I came out. 'She's hungry. I must take her home.' His wife had a day off tomorrow and he was going to visit his 'pit', one of the few authorised moldavite sources. We arranged to meet in the morning.

From the Moldavite Museum I crossed the bridge again, back over the silvery waters of the Vltava river, and up the steps to the castle. Its clifftop façade rose high above the town's roofs, looking down on them like a judge on a courtroom. Two buzzards circled the polychrome tower, their wings stationary, their cruciform shapes black against a heavy sky.

During the time of Rudolf II, Český Krumlov was considered the 'second Prague' and its castle – seat of the Rosenberg family – was the centre of power in southern Bohemia. The Rosenbergs owed their position largely to local deposits of gold and silver, which had brought German miners south from Joachimsthal and alchemists and occultists from everywhere.

Like Rudolf II, the Rosenbergs were involved in the quest for the Philosopher's Stone. Český Krumlov became a centre of esoteric learning, of alchemists, polymaths and pansophists. Well-known names of that ilk survive in the town's scant archival material – Dr Tadeáš Hájek, Bavor Rodovský, Claudius Syrrus, Antonín Michael and Jakub Hořčický (who later became Rudolf II's physician and head of the imperial laboratories). Rudolf himself was a visitor to Peter Vok in his castle. For an extended spell John Dee and Edward Kelley also came under the patronage of Peter Vok of Rosenberg. Mineral extraction and alchemy, the Great Work, were intricately linked – a memo from one of the Rosenbergs discusses both the practicalities of *Bergwerk* – 'mining' and *die andere Heimlichkeit* – 'the other secret' – how gold is generated in the soil.

Worldly reality caught up with the Rosenbergs. Debts grew. Peter Vok was forced to cede Český Krumlov to Rudolf II. Vok vacated the castle and its dozens of rooms and workshops with nothing but twenty-three chests of his cherished books. Rudolf moved his favourite son and heir to Český Krumlov.

In Prague, the emperor's own earthly power was also on the wane, undermined by his involvement with the occult. In 1606, he minted a silver thaler showing himself as an alchemist with magic symbols on his coat. The coin found its way to Rome, where the Pope was appalled. He sent a message warning about dabbling in the 'inferior world'. Rudolf's own brother made it widely known that the emperor had 'abandoned God entirely' and listened only to 'wizards, alchemists, Cabbalists and the like'.

Then, from here, from the halls of Český Krumlov Castle, came another blow: Rudolf heard that his son had been found crazed and naked, clinging to the body of a barber's daughter whom he had murdered and mutilated. One by one, Rudolf was losing his kingdoms: Hungary, Austria, Moravia and finally Bohemia. Early in 1612, his beloved lion died. As Tycho Brahe had predicted many years earlier, Rudolf's own death followed three days later. Within a few years, all his pursuits and beliefs – all his strivings for the substance of unity and truth – were swept aside when religious divisions erupted into one of Europe's bloodiest ever periods of conflict.

Summer clouds sagged overhead. Mid-morning was unusually dark, as though dawn had somehow gone off at half-cock. We were standing in a hole. In front of us was a seeping, head-high cliff of clay.

'Like this.' Vít jabbed his pick at the clay, then inspected what it exposed. He bent to gather small lumps of dislodged material, sifted it with thumb and forefinger. 'You just carry on cutting back. Like this.'

Vít had been granted this bit of land to dig for moldavites – not for direct gain, but as something extra for the visitors to the museum. His own experience had identified it as a likely place and earlier in the spring he'd brought in a digger to make the initial cut.

'Have you had much from it?' I asked.

'Actually no. It's still new.'

Our boots were standing in lace-deep pools of water. I scrutinised each soggy brown lump as it emerged. Nothing. After an hour or more, still nothing. We called it a day. I remembered similar failures from years ago: misery following a fruitless trip to Lyme Regis after reading about Mary Anning; numerous cliffs and outcrops that promised much, and yielded zilch. Finds often came when least expected.

We went instead to see the more clandestine workings. In the forest, it was darker still – but the ground was pale and pitted. There were pits everywhere, dozens – bowl-shaped gouges in the whitish clay about the depth of a person's height, with a top layer of soil and hanging roots. The pits had undermined the trees. Many had fallen to slump against others or criss-cross the forest floor. The whole place felt storm-ravaged.

'They come in from other places now, Prague and further.' Vít was pointing into a ripped-open bowl of soil. 'They just dig – they don't care. The police do little.' Then he looked up, as if aware of being watched. 'Often they are violent.'

The flip side of the gem trade. Conflict diamonds in Sierra Leone, emerald gangs in Colombia, moldavite hunters in southern Bohemia – what sparkles in a jeweller's window does not always have a glittering past.

It started to rain. We could hear it first overhead, a brushing sound that grew louder, and then it was falling on us too. We came to a clearing, a place where the topsoil had been cleared and a large area stripped for drainage. An old digger stood in one corner – its cab windows smashed, its bucket pressed to the earth like a knuckled fist. The ground itself was a swampy scatter of grey ponds and mud.

Suddenly Vít put his arm out to stop me. Below us was a stooping figure. His mud-flecked clothing camouflaged him. He held a short pick and was swinging it, loosening the clay, bending to crumble it in his fingers. He hadn't seen us. When he did, he straightened up and raised the pick. He pulled a balaclava up to cover his face. On it was printed the open-mouthed expression of Munch's *Scream*.

'You alone?' called Vít.

The Scream did not move.

'We . . . it's just us.' Vít gestured at himself and me.

The Scream continued to hold up the pick.

'You found anything?' called Vít.

He remained unmoving. Some weighing up went on behind the mask, and then the pick dropped. He pulled down his face covering, scratched at his chin, half-smiled.

'*Nula*. Not one.'

He had been at it all night. His fingers were automatically breaking the wet clay as he spoke, talking of the rain and how tricky it was in damp ground, and how some nights were like that. With our arrival he was happy to knock off. In a small copse nearby he'd assembled a shelter from willow branches. Plastic bags were heaped in one corner and there was a small stove and a couple of spare picks. He dumped his mud-smeared coat and tools, and as he changed out of his trousers, asked: 'Would you give me a lift?'

I could see Vít was interested in him. This was the side of the moldavite story that he did not know well. His questions were coming fast. They continued as we drove into his village, up to the apartment block where he lived. He invited us in.

So we spent the morning talking, sitting on wobbly deckchairs in the block's basement, its boiler room. The white walls were pocked with black coal marks, the floor stacked with boxes of tools and moldavite paraphernalia.

Vašek warmed up as we spoke. He was full of energy, shooting edgy smiles, carving shapes in the air with his hands, staying still only when it was necessary to light a cigarette. In the flat above were his wife and baby daughter. 'She says keep your work down there. I don't mind what you do there – but don't bring it up here.'

Vašek's life and his business shaped themselves before us. He had been digging moldavites now for seventeen years, since the age of thirteen – he opened his calloused hands as proof. It was his grandfather who got him started. He had worked for the collective farm, digging drains for the fields, and would turn up at their house and take these gems from his pocket and hold them to the light for young Vašek to see.

'The green . . . I remember the colour. Then one day he gave me one. It was my treasure. I wanted to find them myself. I asked him where they were – and he wouldn't tell me. I asked him many times. "If I told you," he said, "you'd never learn."'

His grandfather took Vašek to a place of high ground. 'Well, not that high. We don't have hills round here. My *děda* said, "You must look at the shape of the land, start to see it as a stone-hunter. Then you will find them. It is like learning a language."'

179

Vít chipped in with a memory of his own. 'At the start of my geological training in the Český ráj, the teacher put his hand on a cliff and told us that when this rock sees a proper geologist coming, it talks. When it sees you guys, it's silent.'

Back and forth went the stories – geo-located sites, village names, stone types. Vít answered questions about the museum, Vašek answered questions about his digging and his best finds – the one with a rim like a small ashtray, the perfect teardrop of 26.9 grams, the one with a strange turquoise colour that he found and then dropped and never found again. Now the two of them were scrolling through their phones for images, and each image elicited another, to match and exceed. I watched them: Vít the urban sophisticate trained in media and geology, Vašek from the villages, with no training. Vít wore a 'Moldavite Museum' T-shirt with a simplified diagram of his own design – *15,000,000 years ago – Earth – Asteroid – Bang!* (and an image of a moldavite below). Vašek's T-shirt had tall, elongated letters spelling the word *BEAST*. Each man was devoted to the pursuit of the same quarry, the same stone. It was something more than work.

Vašek was talking about outsiders. He mentioned a group who'd come down recently from Týn to dig. 'They started working in our places. We beat them and chased them off.'

Vít went quiet for a moment.

Vašek stood and left the room. When he came back in, he had a box labelled *MR SPORT GRASS SEED*. 'Look, if I make a hole, I fill it in again – and put down seed.'

A rack of picks stood against the wall. I asked about them. They had clean ash handles and iridescent heat marks. 'I couldn't buy ones that worked well. So I make them myself.'

'From what?'

'I use old lorry suspension springs.'

I remembered a two-day session at a friend's forge in Cornwall, making a knife. We too started with a coiled spring from a vehicle suspension. After hours and hours of hammering and heating, of flattening and graded tempering and more hammering, I had sore shoulders, and the sharpest blade I've ever owned.

Vašek explained the business side. He supplied the moldavites to a man he'd never met, a Czech dealer in the US who sent orders, and Vašek sent back photos and packed them up and the money was transferred. 'Two years ago, it went crazy. I couldn't supply enough. The prices went up and up. Look—' He held up a picture on his phone. It was a customised BMW M2 with an electric-blue paint job. 'I didn't have it for long – the baby came. But for half a year, I had the car.'

It wasn't about the money. 'It gives me enough – but I wouldn't want to do anything else. Even though things go not good sometimes.' He lit another cigarette.

'What do you mean?' I prompted.

'Last winter, there was an accident. My friend runs a pig farm and I knew it was a nice site – I had many good pieces from the land next to his buildings. So I asked him – could I dig from his cellar? I cut through the wall and started a tunnel. I had good stones from that tunnel, the best – big hedgehog-stones. When the tunnel was maybe seven metres long I found a beautiful one. It was round at the end with a pattern on it. It had a long tail – looked like the tail of a cockerel. But the next time I went in, the tunnel collapsed. I was buried. I could not move and my back was twisted.' His voice altered. The trauma of it resurfaced. 'I spent two months in a hospital bed.'

'Didn't it put you off?'

'I thought, it is the end. Now I will work as a plumber or a builder.' He spread his arms around the boiler room. 'But this – it is all I want to do.'

Later, out on the street, we said goodbye. We all had that exhilarated glow that comes from a big talk.

'Wait.' Vašek scurried off inside. When he came back, he held his palm out and in it was a moldavite in the rough form of a maple leaf. 'For you.'

'No,' I said. 'It's your work.'

'I have hundreds. Please.'

And now it sits on the windowsill by my desk. I have not felt its physical energy nor experienced the life changes from this 'stone of transformation', this 'tumultuous tektite'. But sometimes I take it out and look at it up close, using an old magnifying loupe. I try to put a name to its tone, the particular green. I think of the moment of its conception, the heat of the meteor and the arc of silica drops rising into space, and I follow its deeply eroded surface, a mini-Karakorum of mountain ridges and valleys.

Did Rudolf II ever handle a moldavite? It's hard to think he did not. He fastened on any geological curio, any piece of natural magic. He was known to have taken control of the trade of another local gem, the Bohemian garnet. And he came personally to this region, to visit the Rosenbergs of Český Krumlov. But the closest the surviving records reveal is from Anselmus de Boodt, who mentioned 'Czech emeralds'.

After his death, the best of Rudolf's *Kunstkammer* was removed to Vienna. His rocks, gems and crystals, his precious samples of the earth's primal materials, were shoved into barrels and sold. Much of the rest was looted when the Swedes sacked Prague in 1648. The centrepiece of his collection, the table with its magnificent *pietre dure* top, found its way to the Palace of Brussels, where it was destroyed in a fire in 1731.

From Český Krumlov, I took a train across the border to Vienna and on a wet afternoon joined the summer crowds queuing for the Kunsthistorisches Museum. The first floor has a display of the part of Rudolf's *Kunst-* and *Wunderkammer* that was creamed off after his death. A large part is lapidary. There are ornate cups carved from jade and agate and onyx, glyptic mosaics, and *Handsteins* – pieces of natural ore elaborated into mini-tableaux of biblical scenes (one from Jáchymov tells the story of King David and Bathsheba). Each of them was an artistic by-product of the great sixteenth-century Bohemian mining boom, which generated wealth and cosmic enquiry in equal measure, and each one took the more beautiful of the earth's materials and transformed them.

Somewhere in his collection were moldavites. Rudolf and his Hermetic proto-scientists knew nothing of their celestial origins. But it didn't matter. For all jewels were earthly versions of the flickering lights in the night sky, and all stars were gems, and all things were reflected above and below. Beneath our feet, if we look closely, are shards of heaven, and overhead are wonders too and everywhere is the ceaseless dance of the universe – endlessly moving, endlessly changing, endlessly mysterious.

8

MERCURY

ENABLER, DESTROYER, MEDICINE, POISON – slippery spy of the earth's substances. Almost any metal that comes into contact with mercury will disappear, secreted away somewhere inside its liquid chemistry. Mercury is the only metal that flows at room temperature. All the way down to −39 °C it will resist the paralysis of the solid state and remain fluid, remain free. Such is its surface tension that in small quantities it sits perfect, a soft sphere of silvery symmetry. Larger globs of it are pressed to the shape of cushions by gravity. If disturbed, they split into pieces to roll across the surface, a pack of bagatelle balls in skittering motion, and if those pieces meet, they do not bounce off each other but instead fuse in a brief and bulbous wobble and carry on enlarged. Trickster, smiling conjurer – mercury has found ways to subvert the universe's rules, and pose as guide in our ageless quest to discover them.

Of all metals, of all substances, mercury is the one that promised a path to the hidden world. It was the stuff of alchemists and mystics. One of the Sanskrit words for alchemy is *Rasasiddhi* or 'knowledge of mercury'. Early enquirers in Alexandria, like Maria the Jewess, Cleopatra the Alchemist, Zosimos of Panopolis – and later, from the Arab world, Jabir ibn Hayyan – all worked with mercury, seeing in its movement a flashing glimpse of that longed-for place which careful study and crucible work would one day open up. Johannes Kepler, Rudolf II's imperial mathematician and one of Europe's greatest astronomers, had his manuscripts analysed recently, and they revealed in their fibres 'very significant' traces of gold, lead and mercury – suggesting he was not unacquainted with the Hermetic arts.

God of trade and travel, God of eloquence and ideas, God of all things that move without friction through the world. To the Romans, Mercury was the messenger – wings on his heels, bringer of news. He was the fleet psychopomp who leads souls on their journey to the underworld. In Jean Cocteau's 1950 film *Orphée*, the main character reaches Hades through a wall of liquid mercury. The name 'Mercury' was given to that brilliant member of the night sky, the planet that moves swiftest through the heavens, which is coupled by association with its equivalent on earth, the fast-footed orb of metal.

In the fourth century BCE, Theophrastus wrote on just about everything in nature, and he explained how to obtain mercury. First, find cinnabar rock and if it be high up a cliff, you must fire arrows to dislodge it, then take a copper mortar and just a few grains of the reddish stone and with a copper pestle grind it, add a little vinegar, grind some more – and watch. Like manna from the basket of heavenly matter it appears – tiny drops of the miraculous metal.

Insights came to those who contemplated mercury's physical form. To the Greeks, it was the body *soma* while sulphur was the spirit *pneuma*. In the giant cauldron of the earth these two elements combined endlessly, blending into a perpetual range of compounds. 'The metals,' wrote Jabir ibn Hayyan in the ninth century, 'are all of the substance of mercury coagulated with the mineral sulphur that rises in it in a smoky exhalation of the earth.'

That was alchemy – science before science, chemistry before chemistry, the mystery of matter contained in the mystery of metals. In mercury's patterns are the universal patterns – of all movement and all growth and the mystery of life itself. The English name 'quicksilver' does not mean 'quick' as in 'fast', but 'quick' as in 'alive', as in living rock.

If mercury was alive, why should it not also prolong life? Emperor Qin was the first ruler of China and when he wanted to extend his own life, his alchemists gave him elixirs of mercury. Qin's challenge to mortality rather backfired, but his mausoleum did not. Its compound was some twenty square miles in area and guarded by the eight-thousand-strong Terracotta Army. The central tomb remains unopened to this day but is said to contain a model, a miniaturised China, with the jewel-studded ceiling the sky, and the lands below crossed by rivers of mercury.

For 3,500 years Chinese medicine has used mercury – for jangling nerves, cerebral apoplexy, bacterial infection, swelling, convulsions, damaged tissue. The word for 'poison' in Chinese – 'du' – also suggests potency and healing. 'The paradox of Du,' explains medical historian Yan Liu, 'lies in its entwined potential of benefit and harm.' 'Poison' in German is das Gift, originally meaning 'something given'.

In Renaissance Europe, alchemist and mystic Paracelsus pioneered the science of toxicology. He believed disease was caused by an imbalance in the body between the tria prima – sulphur, mercury and salt. He recommended mercury to cure syphilis, which remained the standard therapy until the twentieth century. Young blades were warned – 'Una horae cum Venere et decem anni cum Mercurio' – 'one hour with Venus and ten years with Mercury'; the nursery rhyme 'Rub-a-dub-dub / Three men in a tub' references the immersive syphilis treatment. In his pre-presidency years, Abraham Lincoln relied on mercury pills to treat his depression.

Mercury's dangers were always suspected. Miners developed the 'shakes'. Mercury used by milliners to fix felt made them 'mad hatters'. Many patients treated with mercury showed alarming side effects – they salivated, they lost teeth, they grew red in the face. Robert Burns, Franz Schubert, Charlotte Brontë and Louisa May Alcott are just a few of the better-known names from the late eighteenth and the nineteenth centuries who suffered after being prescribed mercury. It is estimated that Lincoln's daily dose of mercury was nine thousand times above the safety level. He stopped the pills when he became president because they made him 'cross'; American history might have taken a different course had he continued taking mercury.

Minamata is a seaside town in Japan. At one time, the main industry was the factory run by Chisso Corporation, the most advanced chemical works in Japan. From the 1930s, it produced a range of substances – acetylene, acetic acid, vinyl chloride and octanol, all instrumental in building the edifice of Japanese industry. But in 1956, it was noticed that cats in the town were behaving strangely. They displayed symptoms of what was called 'dancing cat disease'. Then Minamata's human population began to display similar symptoms – ataxia, disturbed gait and speech, impaired vision, numbness in feet and hands, insanity, coma, death.

The chemical transformations performed at the Chisso factory used mercury, large amounts of it, and it was seeping out through wastewater into the bay. There it turned into the lethal organic compound methylmercury. In that guise it climbed up the food chain to lodge in the flesh of shellfish – the main diet not only of Minamata's people but, through tossings-away and scraps, the town's cats.

The mid-twentieth century had already seen the catastrophic

effects of radium and radioactivity. Minamata disease coincided with the beginnings of a more general realisation that the earth's resources might appear beneficial, but using them at scale had delayed effects that were often unexpected and alarming, and that companies like Chisso might not always be candid about unintended consequences.

In the 1960s and 1970s, awareness gathered pace. More and more studies revealed profound damage from the use of chemicals. Yet many practices accelerated. In Iraq in 1971, thousands of people were poisoned by eating seed grain treated with a methylmercury fungicide. It took until 2017 for the Minamata Convention on Mercury Poisoning to be adopted. And still it goes on. Between ten and nineteen million people worldwide operate in individual or small-scale gold operations, accounting for about a fifth of annual gold production. Mercury is used to extract pure gold, exposing workers to its toxins while residues seep out into the water supply.

The gods set a honeytrap when they placed mercury in the earth. Its silver globules are endlessly fascinating. In small quantities it works like medicine while its strange properties promise secret knowledge. In Renaissance Europe, the silver liquid was in high demand. So when in 1490 a man saw pellets of it in his bucket at a Slovenian spring, the news spread quickly. Within a short time, prospectors showed up. A small town bloomed. It was named Idrija. Miners brought tools and made holes in the ground; smelters constructed pyres for processing. Over the coming centuries the pyres were improved, the holes became mines and the burrowing went further and further into the rock. Idrija's lodes of native mercury and cinnabar showed no signs of running out. From them has come 13% of all the mercury ever taken from the earth.

From Vienna a bus to Slovenia, and from Ljubljana a smaller bus over the hills to Idrija. The road dropped from the plateau, suddenly, coiling into hairpins that looped and switched and doubled back as they disappeared into mist and rain. For centuries the only way to Idrija was by mountain path. The town existed in isolation in a steep-sided basin, sending out its precious fluid in small flasks, slotted into saddle-packs, ready to transform all it encountered.

Bojan was at the bus station, all smiles and waves. 'How was your journey? Wet? Floods in the east, not so bad here.'

We walked across town to a bar. I propped my bag in the corner, and we sat and ordered beer.

'So,' said Bojan, chinking my glass. '*Srečno!*'

'*Srečno!*' It had been a long day travelling. I drank deep. 'It's good to be here.'

Finding Bojan had not been that hard. A brief googling had thrown up a stack of information about Idrija and some academic articles about mercury's toxicity that occupied me for a while, and then the website of the Idrija UNESCO Global Geopark – 'an area with extraordinary geological and other natural characteristics and landmarks'. I scrolled down until I reached the email address.

'Mr Philip,' began the reply, 'it would be an honour to explain "mercury tears wally". Let me know when you'll be here.' And he had signed off, 'Greetings from Idrija with miners greeting – SREČNO!'

'What did you mean by "mercury tears wally"?' I now asked.

'Just my expression!' He chuckled. 'How I think of mercury. Hg tears, mercury tears – all the suffering.'

When Slovenia was still in Yugoslavia, Bojan had been called in from Croatia. His task was not the tracking of lodes, pursuing ore deeper into the rock, but the reverse. He was put in charge of closing down Idrija's mines.

'Mercury mining was at last coming to a close. If we'd just walked away here in Idrija, the Carboniferous shales would have collapsed as soon as they became wet.' He held his hand flat above the table. 'The town would have fallen into the hole. Bang!' His palm slapped down. 'We have a very difficult geology here.'

Bojan had become something of a figure in Idrija. He had been deputy mayor. He had represented the region in the country's parliament. He had been instrumental in winning the UNESCO Geopark designation. But more than any of that, Bojan was a geologist. At university he had started studying medicine but discovered his vocation when he switched to rocks. He loved the earth's minerals and their mixing and saw in them a snapshot of the drama that is the earth's real rhythm. He used the geologist's excuse when arriving late: 'What's thirty minutes in geological time?' And over the coming days I noticed how his conversation – always fluent and engaging – became more animate when wrapped around words like 'aulacogens', 'clastites' and 'syngenetic ores'.

Bojan quoted, with not a little pride, the reflections of a Soviet geologist: 'Vladimir Smirnov had visited many of the world's geological sites, but when he came here, he said, "I openly admit that such a complex geological structure as in Idrija, I haven't yet seen."'

Soon after he arrived in Idrija and assessed the problem, Bojan devised a scheme to stabilise the honeycombed ground. He injected concrete into the mine's workings – not all of them, but just those that were most unstable.

'In Cornwall,' I told him, 'most of the mines were just abandoned overnight.'

'You have stable rocks,' he said. 'Here, it took twelve years – and a lot of concrete.' We finished our beer and walked out into the main square.

'We will meet tomorrow at Anthony Shaft,' Bojan told me. 'You don't mind going down?'

'It's why I've come.'

With pick, with shovel, with sledge and chisel, with auger and fuse and powder, with pneumatic hammer drill, the miners dug and gouged and the waste fell in slabs and boulders and dust. In half a millennium of chasing mercury and its ore, miners created over four hundred miles of workings below Idrija. There were twenty levels in all, down to a depth of 382 metres. The lower levels, where Bojan's concrete fillings had stabilised the rock, are now flooded.

It took us – I counted – thirty-three ladders just to reach the last accessible level. We stepped from the bottom rung and into a tunnel fork. Darkness went off in three directions, and the darkness was the same shape in each one.

'This way,' said Bojan. He knew.

Our helmet lamps yellowed the rock ahead; the dust-filled cones of light swayed as we walked into the gape. The ground underfoot was soft, the mine-air thick and heavy and warm. Bojan was keeping up a commentary, waving his arms this way and that. 'Pyrite lens . . . fault-line . . . thrusting . . . conglomerate . . . disco dance—'

'Disco dance?'

He pointed to the rainbow colours of a ferrous seepage. 'My name for it . . . Oh and here. Look in there.'

I bent into a small niche. My light bounced bright off the rock; for a moment I made out nothing. Then, lining the inside of a crevice – like tiny silver eggs in a child's Easter egg hunt – droplets of native mercury. Wo-oh! They looked artificial, placed there by some human hand. But they were entirely natural. I peered in closer. They varied in size, from beads to drawing-pinheads to dots, but against the dark rock behind them they shone. They were very discreet, and very surprising.

'Native mercury is rare,' explained Bojan. 'It is one of the unusual things about the Idrija deposits. We have both cinnabar ore and native mercury. That doesn't happen in other places.'

The morning's subterranean walk was filled with Bojan's facts and anecdotes and his enthusiasm for every shift in the walls' texture. He identified the place where the dolomites met the conglomerate. 'If that becomes wet, it will shear.' Here was his favourite bench. 'If I had a problem when working, or perhaps I was angry – my wife has another word for it – I came down here and turned off my light and sat in total black. It always worked.'

The solid rock of the tunnel had given way to softer rock. The walls were lined in this section with countless lengths of softwood. Mining always required a lot of wood. In Idrija, they had a whole system of dams up in the high forests and they would chop and strip the trees and fill the reservoir with logs and then open the dams and sluice them all down in a giant bumping woody cascade until they reached a pool below the mine.

We came to a place where a candyfloss of white fluff billowed from the rock.

'The miners called it "the beard of Perkmandic" – he was the elf that they thought lived down here. Magnesium sulphate. Taste it.'

I put it to my tongue. Bitter. Epsom salts are also made up of magnesium sulphate. They were used as a pick-me-up. The effects have never been medically proved, but it's known that many people have a magnesium deficiency.

'The miners would come down here on Monday. Maybe they had been drinking at the weekend, feeling a bit ill. A mouthful of this – and they are strong!'

In places on the rock face, it had formed into filaments, clusters of long crystals. They looked solid, like tiny pipettes. When I touched them, they crumbled to dust.

The mine was full of little wonders, an Aladdin's cave of mineral tricks. But the one that stayed, that kept coming back to me all morning, and for all the days I spent in Idrija, was those beadlets of mercury – as if the living earth was sweating silver, as if all the exertion of just being solid on a molten mantle produced these moist little gems on the planet's brow. How did it do that? Every time I saw such things, I was aware of a feeling of awe deep in my stomach.

Just a few decades after Idrija's miners began mining for mercury, they had a visitor. Philippus Aureolus Theophrastus Bombastus von Hohenheim – the German-speaking Swiss better known as Paracelsus – was already well travelled. Early in the sixteenth century, there were not many who could claim to have been to England, France, Germany, Granada, Venice and Rhodes, or to have mixed with Cossacks on Russia's southern steppe, served as a 'barber-surgeon' in Dutch and Italian wars and escorted a Tartar prince from Moscow to Constantinople in order to visit a magician. But travel was essential to Paracelsus's beliefs, the only truly empirical way to understand the full extent of the world's marvels.

'A doctor,' he wrote, 'must seek out old wives, gypsies, sorcerers, wandering tribes, old robbers, and such outlaws and take lessons from them. We must seek for ourselves, travel through the countries, and experience much, and when we have experienced all sorts of things, we must hold fast that which is good.'

Wandering medic, book-torching humanist, committed alchemist and frequenter of low places, the Luther of medicine, cultish prophet – Paracelsus was one of the most colourful figures of the Renaissance. His radical thinking did much to shake up theocentric Europe, to place nature in all its complexity on the altar of human attention.

It had long been believed by Western and Arab alchemists that the earth's two elemental constituents were sulphur and mercury. To these Paracelsus added salt to make up the *tria prima*. Of the three, mercury was the most elusive, the most enigmatic, the one that represented the forces of process and change. Paracelsus had come to Idrija's remote forest station to watch the transformation of mercury itself – the moment it was delivered from the womb of the earth. He had always sought out mines and miners on his travels; he'd conducted experimental work with a range of ores and substances and stones. He was the first to observe hydrogen and nitrogen; he gave the name to zinc. He dismissed as futile and misguided the idea of chrysopoeia, the transmutation of base metals to gold, but believed strongly in the broader principles of alchemy – universal pattern, the links between the human body and the minerals beneath our feet, between botany and geology, between the forces of the earth and those of the heavens.

In Idrija he saw mercury in its native form, and he saw the blush of cinnabar ore. He saw the processing – mule trains that carried mined ore up into the forest to be superheated in giant hearths. And he saw the result: liquid poured from clay retorts into skins, a hard-won metallic slick which flowed in such a singular fashion that it suggested the essence of creation.

But Paracelsus also saw something else. Men who were not working, who sat outside their homes shaking, or lay in bed coughing, vomiting, unable to move their limbs. He saw the missing teeth, skin pocked and colourless. 'Everyone that lives here,' he wrote, 'is bent and paralyzed, partly asthmatic and partly chilled through.' Something unseen and malign was happening alongside the extraction, the digging and the processing. That, he was told, was the work of rock-spirits, the wrath of Perkmandic.

Paracelsus was used to questioning received notions, particularly when they involved spirits. What was happening in Idrija, he proposed, was not the work of devils. It was what happened with uncontrolled exposure to strong chemicals. He recommended breathing filters for the mercury miners (a measure which was not fully adopted until over four centuries later in the 1950s). He believed in the therapeutic properties of mercury but also knew its dangers; achieving a balance was at the core of his clinical thinking. He rejected the prevalent medical idea, spurred by the recent rediscovery of the work of Galen, that disease must be treated by rebalancing the four humours. His view was that ailments needed precisely locating, then treating directly. Organs did particular jobs, and targeted toxins could assist in ridding the body of disease. 'The dose makes the poison' is the adage widely used to distil his principle of moderation. He was alone in his views and banished by the medical establishment. But he is now known for introducing chemistry to medicine, celebrated as the 'Herald of Toxicology' and the 'Godfather of Chemotherapy'.

Paracelsus's interests and ideas went beyond medicine. His calling was to address the entire question of nature and our place in it, the great humanist adventure of exploring the cosmos without the guidance of God. Humans, he believed, were part of the natural world, not divinely licensed agents standing outside it: 'There is nothing in Heaven and on earth which is not in man. And God, who is in Heaven, is in man.' He did ascribe certain faculties of God to humans – and among them was the imagination, the ability to see patterns behind the visible world: 'He who is born in imagination discovers the latent forces of Nature . . . Besides the stars that are established, there is yet another – Imagination – that begets a new star and a new heaven.'

Over the coming centuries, Paracelsus spawned many followers, and many and varied schools of thought. To the Rosicrucians, he was a prophet, to homeopaths an early advocate. The great German mystic Jakob Böhme echoed Paracelsus and his notion of the divine in us all when he cried: 'Where is God? Listen, you blind human, you live in God and God in you.' For Goethe, Paracelsus was a visionary and a model for his *Faust*. When William Blake spoke of his influences, he mentioned the moment Paracelsus and Böhme both 'appeared to him'; he elevated the visionary element in their work above even Dante and Shakespeare. For Blake too, the imagination was the spark of God in us all, the unique ability to see unity and truth behind the physical. 'The imagination,' he wrote, 'is not a state: it is the human existence itself.' In his book *Jerusalem: The Emanation of the Great Albion*, he put it: 'Imagination is the real and eternal world of which this vegetable universe is but a faint shadow.'

When Paracelsus came to Idrija to watch mercury emerge from the earth, he witnessed something that went beyond this world. He saw a chemical messenger from a place not normally visible. Look down on a pool of mercury. Watch its silvery surface and what it reflects: your own face and the sky above. Everything, he wrote, is a 'microcosm, a condensation of the entire universe'.

*

With Bojan, I went to meet a retired mercury miner. Andrej had gone down the pit at fifteen and spent thirty years working there – until it closed in the 1990s. He was standing among his vegetables. Marta, his wife, stepped out of a small polytunnel, and Bojan greeted them both. There was a brief conversation about potatoes.

'I had four hundred kilos last year,' said Bojan. 'I could hardly give them away. But they were big.'

'Ours too were big. We went to dig up the new ones and they were already like this.' Marta shaped her hands to something between an orange and a grapefruit. She said this year she was experimenting with some new tomatoes, and Bojan mentioned his chosen varieties, and Andrej began to talk about beans. The vegetable talk went on for some time.

Marta and Andrej lived in a flat in a sturdy building constructed between the wars when the Italians ruled this part of Slovenia. It had deep eaves and green louvre shutters. We were delayed further because Bojan bent to examine some fossils in the stonework by the front door. 'Triassic limestone, chert – a lot of small shells.'

Andrej's restoration of their first-floor flat meant following boards across open joists. Clusters of wires hung from the ceiling. But in the kitchen, everything was in order. Marta pulled down a jar of home-made yoghurt. She laid out saucers of home-made compote. She poured glasses of home-made lemonade. With the table full and all of us spooning up fruit jam, she smiled and sat down.

'We felt the effects, yes.' Andrej's meaty forearms lay heavy on the oilcloth. 'Waking up in the night shaking and sweating . . . mouth full of saliva.' He shrugged. 'We'd take maybe a day or two off.' He looked at me with a half-amused expression.

Bojan too had felt it. 'Tremors, a sweet taste in the mouth. I've lost some teeth.'

But they didn't want to talk too much about it. What Andrej wanted to get across was the *esprit de corps*. 'It was hard work but everyone was friendly. We were all *tovariši*, comrades.'

He missed it. Here was I asking about poisoning, prying into his health, and all the time he was longing for a way of life that was now gone, for Tito's Yugoslavia, for the working town with its thirty-two bars ('walking home it was impossible to avoid them'), for his fellow workers – friends, cousins, even those who irritated him. 'They all had nicknames – usually the region they came from.' He remembered 'Wind Man' (the air monitor), 'Fine Ore' (an engineer), 'Golden Retriever' (the one who brought sausages to work). They all took their tasks seriously but when it was over, they knew how to relax. They had a code for tapping the pipes: '*tsek*', a short sound that meant 'bring down a bottle'.

'It was like Las Vegas,' Andrej grinned. 'What happens down the mine stays down the mine. We were frightened of one thing only.'

'What was that?'

'The safety crew. We had to hide the bottles when Tatjana came down.'

'Sorry it is not so tidy!'

In the main office of the Mercury Heritage Management Centre, Tatjana Dizdarevic moved her heater from behind a chair. 'We have to keep the windows open for ventilation. Radon. I am moving offices, so everything – well, look! – it's spread all over.' She cleared some files from the table, and we sat down opposite each other.

Tatjana did not look frightening. She had a kind and scholarly look. But for twenty-five years she'd been the mine's Head of Department of Safety at Work. Her role had been to protect men working in an intrinsically dangerous place – to resist any tendency to cavalier practice. Some tension was inevitable.

Even without a drop of mercury being produced, Idrija remains highly toxic. Thirty thousand tonnes of mercury remains around the smelting sites and the dumps. The banks of the river, where processing had taken place for centuries, are full of mercury. It finds its way down into the Gulf of Trieste. During times of heavy rainfall – like now – levels of dissolved and particulate mercury rise significantly around Venice. Mercury extraction continues to call in its debts. 'We must always monitor. It never stops.'

I asked how she felt about mercury now.

She sat back. 'From a scientific perspective, it is fascinating. I never tire of studying it. But you must handle it properly. You must be respectful.'

As a girl, Tatjana remembered playing with mercury in the streets. She lived near the old smelting plant. 'We loved the little drops. We liked to watch them roll around in our hands.'

After school, she tried for one of the scholarships offered by the mine. 'They were a bit surprised because girls didn't apply. They said that women are allowed to do anything except work underground. So I went into safety.'

She worked her way up to become head of her department. Since the mine closed, she has continued to keep an eye on the mine. She has become involved more widely with the question of toxicity and mining, publishing academic articles and joining the United Nations Environment Programme.

Some years ago, she travelled to Kyrgyzstan to visit its last mercury mine.

'We had come to try and close it down. The director was polite but of course he wasn't very pleased to see us. He told me I was wrong – mercury is not poisonous. I watched him scoop up a handful of mercury from a bowl. And he drank it! Apart from that, they were nice people, very kind.'

Like Jáchymov, Idrija is a place that presents the most potent of nature's forces, life and death, in elemental form. I found the atmosphere heavy. In the afternoons, I stole away from the town, from the redundant mine and toxic tailings, from the hiss and tumble of the river in its culvert, to walk in the surrounding hills and forests. Sometimes the path followed a wooded ridge, steep as a roof on each side, with views of more valleys through the leaves, more forested ridges. I loved those forests, in full bloom – the delicate hornbeam, maple leaves outsize and flopping, thin young beeches rising like the notes of a song into the canopy. In the meadows were orchids, and everywhere a multicoloured flora of columbines and wild strawberries and campanula.

The same complicated nexus of geology that produces Idrija's mercury has created a botanical hotspot. Rich mineral soils and remoteness combine to produce a diversity that now, in the summer, was almost audible. The area is full of curious hybrids and endemics – the snake pine (a single genetic variant with serpentine branches), a white Daphne, the perennial *Hladnikia pastinacifolia* found only in an area of a few square miles, a gentian named 'flower of the sweet lady' and the elusive henbane bell (*Scopolia carniolica*), a deep violet flower that – like mercury itself – is both beautiful and highly toxic. It was originally identified by one of Idrija's more notable residents – its first, ill-starred doctor.

Sent to the town in 1754 to treat miners for mercury poisoning, Giovanni Antonio Scopoli spent much of his time wanting to leave. But Empress Maria Theresa ordered him to stay – the mercury mines were vital to the empire, and she needed the miners healthy. His sixteen years here were characterised by misfortune. It started even before he arrived. He and his young family came with nothing because their boat had sunk en route with all their possessions. A few years later, a fire killed his wife and children. For Scopoli, Idrija's roadless basin became a dark and bitter place.

'I am,' he wrote, 'weary, penniless and powerless, hardly able to struggle on, buried here among miners, almost exhausted by the plight of my miserable existence.'

What saved him were these hills, these forests. He immersed himself in their plants and animals and minerals, speciating and scrutinising them, spotting links between them, exploring the uplands for clues to the earth's mysteries. He described himself as a 'poor man smitten by love of wonderful nature'. He studied everything around him. He found solace in his work as a naturalist, just as his fellow naturalists around Europe found solace in his findings. In *The Natural History of Selborne*, Gilbert White mentions Scopoli fifteen times.

Carl Linnaeus was another admirer. Scopoli sent the Swede his own books – on flora, on entomology, on the effects of mercury on the human body. Linnaeus recognised Scopoli's dedication and talent: 'Your letter contains observations so rich that they would persuade even the Gods . . . if only we had more Scopolis, the natural history of animals would soon reach its apex.' They kept up a correspondence for years.

Scopoli shared with Linnaeus not just his writing, but also seeds and plants and bones and butterflies, and he sent him rocks and notes on rocks (the actual samples tended to get lost in transit). Both men's love of nature was for nature as a whole. Each pondered deep on everything – including the processes that led to geological diversity. Linnaeus for his part collected rocks and had long been a visitor to mines; like a surgical dissection, mines offered the observer the chance to see something of the earth's organs. Linnaeus retained the alchemists' idea that underground 'ores grew gradually'. He suggested they were conceived in a similar way to living things by a process akin to sexual reproduction. Certain liquids (male) combined with certain soils (female) to produce rocks in all their different forms. His ideas on mineralogy haven't had quite the impact as his organic taxonomy.

But when it came to minerals, the mid-eighteenth century was still a time for open speculation, unburdened as it was by too much detail. James Hutton had yet to deliver the 1788 paper on long-term erosion, seen as the launch of modern geology. James Ussher's six-thousand-year timeline remained the consensus for the earth's age. Alchemy was still in the room. Neptunists and Plutonists bickered over their theories (each as wrong as the other) and Goethe and others were able to toss their own poetic ideas into the mix. There was diversity in the theorising, in the use of analogy and metaphor, and there was joy in the joining of disparate elements. The assumption was of commonality between all aspects of the natural world, and Linnaeus's work linked as well as divided. The playwright August Strindberg said of him that he 'was in reality a poet who happened to become a naturalist'.

Linnaeus never met his protégé Scopoli, but he honoured him by naming a genus of plants after him. *Scopolia* includes the local Idrija species *Scopolia carniolica*, the henbane bell. The flower is now known to be the source of a curious neurotoxin, which also bears the naturalist's name – scopolamine, or Devil's Breath. Hallucinations, coma and death are the result of ingestion but in small quantities it has medicinal benefits – for motion sickness and gastrointestinal spasm and as a pre-med. It also has the effect of breaking down inhibitions and was used as a truth drug by the Czechoslovak secret police during the interrogation of dissidents.

Despite hours of looking, I failed to find the plant. I asked Bojan and he took me to several known sites in the hills. No luck. 'It is not easy finding *carniolica*!' he exclaimed, as if it were the Golden Fleece. 'Each year is more difficult.' Our search though had compensations. We visited a place where there were alpine plants on one slope and Mediterranean ones on the other, and where the dolerite and conglomerates met and formed a spectacular waterfall. We called in on an old friend of Bojan's who had lost his eyesight and now lived deep in the forest with his wife. He'd put in a small swimming pool for his grandchildren, and we drank hazelnut schnapps on his terrace and he said: 'I can still see, but only with my ears.'

On the way back to Idrija, we went to look at a geological site. A narrow goat-path led along the side of a hill. The slope was steep above and steep below. Rockfalls blocked it in places. We came to a knuckle of limestone above the path and suddenly Bojan cried, '*Carniolica* – look!'

We scrambled up. A rosette of green, alone against the rock. It didn't look much. Waxy leaves and a brownish bellflower. But bending down over it, I was aware of its rarity and its hidden powers. I thought of those jewels of native mercury in St Anthony mine, with the same aura.

'That is it,' Bojan confirmed. 'Yes, that is *carniolica*.'

We carried on along the narrow path. Bojan was talking back over his shoulder. 'Some young guys in town tried the root. They wanted to see what happened to them.'

'How did that go?'

'Not well. Not well at all. They are not really the same.'

*

Mercury from Idrija found its first practical use, its first commercial market, in Venetian mirrors. Mercury was the surface in which the burghers of Renaissance Italy – the Medicis and Borgias and Pazzis – admired themselves. Designers used mirrors to increase light in interior space. The process of applying the mercury was highly secretive and dangerous, a closely guarded craft carried out on the Venetian island of Murano; disclosing the method meant execution, if you survived the mercury poisoning.

In 1509, Idrija passed from the Venetians to the Holy Roman Empire and the mercury began to flow north, into the hands of the Habsburgs. At one time, mercury contributed 10% of their entire revenues. Flasks of it arrived in Rudolf II's Prague, where it was used for experimentation, in the endless pursuit of transmutation. But the high price of mercury was maintained neither by mirror-makers nor by alchemists but by the part it played in the 'patio process', in the separation of silver from its ore. Late in the sixteenth century, mercury from Idrija began to be traded through the exchanges and wharves of the Netherlands. From there, it was taken out to the New World, where it transformed Potosí's pyrargyrite to silver pesos, which themselves spread on around the world, across the Pacific, back across the Atlantic, transforming the wishes of emperors and queens into material assets – palaces and armies, roads and ships – which in turn acted on the earth, altering it, helping to initiate the great age of transformation. Mercury was the éminence grise behind it all.

'There are as many mercuries as there are things,' wrote Paracelsus. Mercury performed physical tasks but there was also 'philosophical mercury', a parallel entity that extended the metal into a world of abstraction. Each plant and each animal has its own mercury. Human mercury is vitality, and also the imagination, the faculty that transforms spirit into image, time into story, the past into memory, diversity to unity.

A week ago, I had been reading in the Jan Palach Library in the Old City of Prague. Below the window, the ancient headstones of the Jewish Cemetery were leaning this way and that. It was midday, and it was hot. I was feeling drowsy. Before me was a large book entitled *Alchemy and Rudolf II*. The text was running by me without releasing any of its meaning. I found myself rereading the same paragraphs. I turned the page – and was at once jolted awake by an image. The more I looked at it, the weirder it became. It was from the sixteenth-century Czech translation of the *Rosarium Philosophorum*, one of the most widely circulated alchemical texts in early modern Europe.

Mercury is standing naked. He has on a pair of red wings. On his head is a raven, and on the back of the raven is a white dove and beside the two birds is some sort of crustacean, a crayfish or lobster with one claw on the dove and one on Mercury's hair. A snake is wrapped around Mercury's waist and its tail loops in beneath his genitals. Something blue and fish-like sticks out of his arse. He holds a stick, an arrow and an axe in one hand; in the other is a Bouffadou fire pipe, which he is blowing. One of his feet is cloven, the other piscatorial, and they both rest on the top of a small moon. A skull lies on the ground beside him and Mercury also has a dog on a leash and beside the dog are two figures – man and woman / sun and moon – lying together on the grass, arms crossed chastely across their chests.

The picture was so burdened with meaning it made me smile. The entire canon of alchemical literature is full of surreal and symbolic pictures – bicephalous hermaphrodites, trees growing from human stomachs and a lot of couples having sex in various ways (the great scholar of Hermeticism, Lawrence Principe, is not the only one to draw strong parallels between alchemy and procreation).

What it all means is perhaps less relevant than the sheer range of it, the extent to which mercury is a vehicle for meaning itself. Mercury is central to the *Rosarium Philosophorum* – common mercury and philosophical mercury. It is mentioned two hundred times in the meandering text: '. . . bodies are dissolved into philosophical mercury, that is into water of our mercury, and it is made one new body . . . in Mercury there are but two elements in action, that is to say earth and water . . .'

I returned the book and left the library. Outside, the Prague air was oven-hot. I was restless and stimulated. A group of people stood outside the Jewish Cemetery, semicircled around a guide. 'You've heard of the Golem, right? Rabbi Loew is buried right here, behind this wall. The Golem – no? Well this Loew guy created a living being from the river's clay, a living being . . .'

I walked on. The afternoon sun quivered on the waters of the Vltava. Distant hills faded into haze, which merged with the haze of the sky. I followed the river to where it turned eastwards, past the pleasure boat quays and the double-moored commercial barges, past the city's golf course and the river shacks, and I was happy to walk in the heat among all the riparian muddle and mess and follow the whirl of my library thoughts.

Esoteric literature can have various effects. Sometimes it's deliberately arcane, playing word games to give the illusion of hidden profundity. Sometimes it's plain nonsense. At other times, a sort of spell is cast. Shapes come into being behind the words, beside the words, forming and re-forming. It does not happen if you try and understand too much of what's being said. Naked Mercury and his menagerie and the text of the *Rosarium Philosophorum* had played that trick. I was struck by the way the river fitted its course, and the sun made stars on its surface and the hills made boundaries for the valley and the contrails in the sky suggested the network of roads below. I carried on along the river-bank and let one thought transform into another, and another, and another . . .

Alchemy in practice.

On my last day in Idrija, I drove with Bojan above the town, and we set off on foot through the trees. Beech twigs cracked beneath our boots; sunlight came lancing through the canopy above. Bojan was looking left and right as he walked. He was holding a geological hammer by the head, letting the shaft swing at his thigh, and it looked like a part of him, a happy adjunct to his life of geosearching. We were after the old hearths; hundreds of them were scattered through the forest.

'Wood was always the problem.' Bojan's voice was echoey among the trees. 'To begin with, it was easier to bring the ore up here than take wood down.' We reached a flattish area. 'They would build up a great pile of wood, then cover it in moss and branches.'

I'd seen the diagrams – sectional images of stacked wood, great mounds of felled timber carefully assembled. Placed beneath the wood were dozens of earthenware vessels and each one was filled with ore. The fire was lit. At 600 °C, the ore would start to melt and the mercury flow into the lower section of the vessel.

We dropped down to a gulley. Bojan started digging with his hammer. 'Sometimes you can find pieces . . .'

I dug with my hands and spotted an orange flake. 'Is this something?'

He glanced at it. 'Small piece.'

We carried on. 'Here, look!' called Bojan. 'A part of the throat.'

He pulled out a much larger shard. It had the curve of the vessel, and on the inside was a smudge of red glaze where the cinnabar had re-oxidised. Suddenly, they were everywhere – half-embedded in the soil, their fractured edges rounded by age. Within a short time, our hands were full of them.

'When you see one, they all appear.' Bojan was excited. 'Like mushrooms!'

We carried on, hunched over the ground, picking at it. And the site opened up to a phantom image of its sixteenth-century self. Horses and wood-piles, smoke in the trees, seeping up from the brush-covered clamp, groups of women working – the equivalent of Cornish bal maidens – who did not go underground but toiled at crushing and sorting the ore, the final pouring of mercury into leather flasks.

Smelting always had mystique, a magical process discovered by chance. But it was dangerous magic. Rituals and traditions suggest that it was an art widely believed to be stolen, a Promethean gift that needed propitiation. From central India comes a series of myths that tell of a deity tricking smiths into their own furnaces, to be burnt alive. The Asur are a tribe of metalworkers from the same area, whose forges are thought of as 'death wells'; human sacrifice was said to be part of their smelting traditions. The Chewa of Malawi required a dead foetus to be placed beneath an iron-smelting furnace. A common theme in ancient Iranian, Egyptian and Greek belief was that metals were the flesh and bones of the gods and you tore at their bodies when you extracted them.

In *The Forge and the Crucible*, Mircea Eliade identifies from many mythologies the same sense of collective guilt, unease in the transforming of metal: 'These myths are emphatic in their *hatred of iron* and metallurgy . . . the same negative and pessimistic attitude which is present, for example, in the theory concerning the ages of the world, in which the Iron Age is regarded as the most tragic and most debased.'

Here in the forests above Idrija, the smelting of mercury was wasteful. The ores were very rich, up to 50% mercury, and it didn't matter that so much of it was lost. Tonnes and tonnes of it remain in the soil, as they do in the dumps below, as they do in soils around the world at thousands of artisanal gold workings. Philosophical mercury is a poetic idea, making sense of the mystery of constant alteration and process. Actual mercury is poison.

In the morning, Bojan came with me to the bus station to say goodbye. As we waited, he pointed at the retaining wall and a large crack. 'The old mine is below here. It's making the rock move. I come past it each morning, walking my little dog. When you're walking your dog, you notice things.'

'Is it a problem?'

'OK for now. But in geological time? Maybe not so good.'

A couple of days later, in Sarajevo, I went to see my friend Dr Bojana Mojsov. Born and brought up in Yugoslavia, Bojana had left the country before it collapsed in the 1990s. She moved to New York to study as an Egyptologist, and from there to Cairo, where she lived for years researching, travelling, writing books. Now she'd come back to the Balkans, to Bosnia.

It was early evening. We were sitting in her garden, eating red cherries from a large green-glaze bowl. Beyond the fence the roofs of the city dropped into the Miljacka valley in tiltings of clay tiles. Ottoman minarets rose from them like cypresses.

Over the years, we'd had many conversations about her work. She'd spoken of the beliefs and mythology of dynastic Egypt, about the Osiris cult, and Alexandria before the Arab conquest – but they'd been general conversations. Now our various projects were in alignment. I knew she'd been looking at the link between ancient Egypt and those Hermetic texts that reached Renaissance Europe. I began by asking about Hermes Trismegistus. Was he just a made-up figure by the Greeks in early Christian Alexandria?

'In part, yes – his name is Greek after all, not Egyptian. And the work attributed to him owes not a little to Plato and the Gnostics. But there are a lot of other ideas in the body of work attributed to him, the *Corpus Hermeticum*, that draw on the beliefs of ancient Egypt.' Bojana had lectured on the subject and she spoke in carefully constructed sentences, yet with every step into the subject I watched her enthusiasm grow. 'Whoever first put pen to paper in his name was drawing on much earlier Egyptian traditions. References to the solar trinity, for instance. In some of the books, the teacher is called Tot or Tat – which refers to the ancient Egyptian god of writing and wisdom Thoth. In fact, the entire body of Hermetic work owes a great deal to Thoth – or *Djehuty* in the original version.'

She spoke about the manuscripts themselves, the physical originals of the *Corpus Hermeticum*, and how lucky it was that they survived at all. During the Persian and Arab invasions in the sixth and seventh centuries, they had to be smuggled out of Alexandria, crossing the Mediterranean hidden in rolled-up carpets, then kept for many centuries in Greek monasteries. With the expansion of the Ottomans in the mid-fifteenth century, they were again at risk, and again they were evacuated, this time to Italy.

'The *Corpus Hermeticum*,' she explained, 'was found in a monastery in Macedonia. A monk brought it to Florence in about 1460 and it caught the attention of Cosimo de' Medici, who was assembling a number of texts to challenge the authority of the Church. Cosimo was already working on Plato's *Republic* through his translator, Marsilio Ficino. And he had the *Enneads* of Plotinus too, waiting to be translated. But Cosimo ordered Marsilio to concentrate on the work of Hermes Trismegistus. Imagine! He had some of the most influential works ever written – but he put them on hold in favour of the Hermetic texts. When it was completed, the *Corpus Hermeticum* went into fifteen further editions and was widely translated around Europe. It was a bestseller.'

'Why so popular, do you think?'

'Well, one thing was its antiquity. At the time, there was a tremendous urge to seek out origins. The *Corpus Hermeticum* was seen – believed – to come from the oldest known source, from ancient Egypt.'

Below us, the muezzin had started the evening call to prayer. The first was joined by another, and another, echoing out across the city.

'The *Corpus Hermeticum* is not an easy text,' I said. 'What do you take away from it?'

She selected a cherry from the bowl. 'One thing is an emphasis on revelation – making the invisible visible. That also comes from ancient Egyptian belief. Wisdom cannot be taught – it must be experienced. The physical world is the route to the "seeing beyond".'

I told her about coming across beads of mercury in Idrija's mine and how it changed my entire view of it. 'Paracelsus went to Idrija to see the production of mercury with his own eyes. Then he became the first physician in Europe to administer it.'

We spoke about Paracelsus – his idea of 'alkahest', the universal solvent which could make everything disappear, and how he initiated the prescribing of opium in Europe and claimed to keep it in the pommel of his sword. We moved on to talk about that strand of European mysticism which grew out of Paracelsus and his work, of Goethe and Böhme and Novalis, all of whom owed much to the Hermetic tradition, and then about Bojana's recent trip to Mystras, last seat of the Byzantine court before Greece fell to the Ottomans, and the belief in Renaissance Europe in the *prisca theologia*, the original belief given to humans, which was then muddled up into various doctrines, and about the love of true alchemists for the physical world, for the natural world.

It had grown dark. We'd hardly noticed. 'A lot of talk!' said Bojana.

'It's a thing about alchemy,' I replied. 'It draws you in – then sends you spinning off in about six directions at once, a bit like mercury itself.'

'There's a passage in one of the Hermetic books that says that really the central message should not be talked about. It cannot be explained – only experienced. If I remember rightly, it goes on to speak of the value of silence.'

'We'd better shut up then.'

And we did, and there was only the night around us and the sound of the night, and the distant hum of the city.

9

COPPER

THE MYSTERY OF COPPER LIES in its colour – or rather, colours. It's the chameleon of metals. In its pure form it has tints of brown and gold, orange and red, but none of these quite captures the full complexity of the shade. It's like oak leaves when they've just come out. When it breaks, copper presents yellowish-purple. Cupreous crystals – as in liroconite and azurite – are like the bluest sky and the bluest sea combined. Oxidised, it turns a rich green. Native copper makes artful shapes, the reddish-brownish-goldish gleam forms trees, or teeth, or fish spines, or the regular fletchings of an arrow.

Copper was the first metal to be smelted at scale, the first metal to be dug and gathered from the ground, to be sharpened to a blade, to be produced for fleshing and fighting. Like bronze to come, it had the advantage that it could be used and re-used, melted and reshaped. That's how the human relationship with processed metal begins. Copper proved not just powerful in the hands, and beautiful as adornment, but biddable. Archaeologists talk of the Chalcolithic Age (from the Greek *khalkos*, 'copper'), a transitional period between the stones of the Neolithic and the tin–copper alloy that gave its name to the Bronze Age.

No other metal occurs so readily in its native form. Does that mean you can go around picking it up? No. It's gone. What can be collected has been collected. The world's largest native copper deposits were in the Great Lakes area of North America, where several thousand years BCE, Native Americans began to collect the nuggets of shiny rock. Cold-hammering hardened it. They made blades for knives and adzes, and jewellery. They traded the copper and it spread widely. Early European explorers saw copper pendants hanging from the ears of the indigenous people; they saw bands of it around their wrists and beads of it around their necks. An Algonquin leader gave the French pioneer Samuel de Champlain a piece of copper a foot long.

Some pieces of American native copper were too large to be moved. Even though people lit fires underneath them, and chipped them smooth with stones and antlers, they remained in place – a storehouse of metal 99% pure. One such boulder weighed one and a half tonnes and lay to the south of Lake Superior beside the Ontonagon river. In the early 1830s, Julius Eldred of Detroit heard about the boulder and thought it would make a nice attraction for his hardware store. Twice he sent expeditions upriver to retrieve it; twice they failed. The third time, he took twenty-one men and they chopped down trees and laid rails through the forest and inched it out. Eldred charged people a fee to see the rock and they arrived by the dozen and he did good trade. Then the government came and took it away. They said it was theirs. More recently, the Keweenaw community claimed it as a sacred object and asked for it back but the courts concluded that the community had no rightful claim. It remains in the Smithsonian Institute in Washington.

Life is impossible without copper. Virtually every living cell requires it. Plants wilt from its absence. Copper deficiency in ruminants causes weakness in the limbs, and their back legs cease to work and they stagger about, a condition known as 'swayback' in sheep and 'falling disease' in cattle. The heavily processed diets of industrialised countries are low in copper. It has particular importance in the brain, enabling the development of neurotransmitters and as a structural component. The misfunctioning of copper is implicated in many neurological disorders – Huntingdon's, Menkes, Wilson's, Parkinson's and Alzheimer's diseases.

Copper is life-giving, but too much is bad for you. Usually the body deals with excess copper, excreting it in bile, but liver malfunction can mean that it accumulates. It is more toxic when mixed with acid. Copper is also poisonous to micro-organisms like bacteria. Doorknobs of brass (a copper alloy) are often preferred in hospitals, because they self-disinfect. Naval ships of the eighteenth century kept out shipworm by sheathing the hulls in copperplate. It was immensely expensive and highly effective but it had a knock-on effect: galvanic corrosion on the ship's iron fastenings led to a number of dramatic sinkings. The navy then replaced every iron nail with copper. Water has long been carried through copper pipes, which are naturally sanitised and do not corrode. Electrical power would not have had such a transformative effect on the planet without the advantages of copper wiring.

Copper is the real messenger metal. Unlike mercury, which actually delivers nothing, copper's conductivity has transmitted countless messages. By 1901, the world's continents were connected by copper cables. A dozen or more crossed the Atlantic and a dense cluster passed up the Red Sea into the Mediterranean. Even before then, copper had long disseminated words through printing. It was used in early inks – copper is one of the metals that make the letters of Gutenberg's Bible glitter in the light. Copper-based dyes proved the clearest for type and the longest lasting. Copper plates were best for printing – soft enough to create a subtle line but hard enough to produce a thousand copies or more.

Until photography, copperplate images were the means by which people learnt of the world beyond their ken – the shapes of exotic animals, distant cities, mountains and jungles, battles, temples and ships of the line, the latest advances in science and industry. Maps and maritime charts enabled European expansion and these too were printed from copper templates. When corrections were needed, and in an era of exploration they constantly were, the relevant part of the plate could be hammered smooth and new details engraved. Copper is one metal driving the new age of nanotechnology, as it drove the industrial age beforehand, as it drove the first great leap forward, several thousand years ago in the Bronze Age. The global transition to renewable energy will soon require twice as much copper per year as was used in the last hundred.

From the Austrian Alps comes evidence of Europe's earliest known 'copper factory'. Between the sixteenth and thirteenth centuries BCE – the height of the Bronze Age – about twenty thousand tonnes of copper was mined and processed in the region of Mitterberg. There were other sources of Bronze Age copper in Europe – the Balkans and Cyprus, Spain, Britain and Ireland – but high in the Austrian Alps, at Mitterberg, was the first known place where it was produced on an industrial scale.

On a warm summer evening, I left the mountain bus and watched it disappear back down the valley. Cowbells sounded from the opposite slope. The low pylons of a ski lift ran up a strip of pasture. I crossed the road to a large, steep-roofed building. In winter it was a ski hostel; now in July it had been commandeered by a team of archaeologists, archaeo-metallurgists, dendrochron-ologists, specialist photographers and students. They had come from northern Germany and, as they had done each summer for the last two decades, were carrying out intensive fieldwork in the Bronze Age copper area of Mitterberg.

I'd met them in the spring. From the Biesbosch, I'd taken a quick detour over the border to Germany, to Bochum and the Deutsches Bergbau-Museum – 'the world's largest interactive mining museum'. Next to the museum was the Institute of Mining Archaeology, and in its library I first came across reports of the Mitterberg copper works. Such large-scale production of copper in the Bronze Age was another clue to the extent of the metal markets and connectedness at that time. Mitterberg copper has been found in hoards from the west of Ireland to the Greek islands, from Scandinavia to the east-ern Danube. It was Mitterberg copper that helped drive Europe's Bronze Age and it was Mitterberg copper, along with Cornish tin and Cornish gold, that was used for the Nebra Sky Disc.

The reports were written in the institute, and soon I was sitting around a table with the authors – Peter, Eva, Jeni and others, all of them preparing for the upcoming field trip to Austria. Here in the industrial heartland of northern Germany, I had the impression that the month in the Alps was a high point of their year.

Peter was sitting at the head of the table, flicking through a bound report, largely written by him, skipping pages and pages of statistical analysis. 'Boring . . . boring . . . boring . . . Ah!' He spread out a map of the Mitterberg area. 'So – you have the main mine over here at Arthurstollen. And in this place here are the dressing and beneficiation sites. They are quite scattered. From about 1700 BCE we see the sudden appearance of settlements in this area. We think that's when the production would have started.'

Peter was a tall man with an appealing vigour and a dark beard. He was wearing a black hoodie and splashed across its back was BEHEMOTH (a Polish extreme metal band). From a shelf, he took a lump of rock and handed it to me.

'You know what that is?'

'Chalcopyrite?' I suggested.

'From Mitterberg. Traces of nickel and arsenic only, in gersdorffite.'

I turned the rock over. 'We have a lot of this in Cornwall.'

'Not with this purity.' Peter took the rock and held it before him, as if it were evidence in a trial. 'One thing about Mitterberg copper is that it was very, very pure.'

We carried on talking, about how the mine was re-discovered in 1830 when copper miners broke through to the prehistoric galleries, and how each year now more Mitterberg copper is discovered around Europe, and about the beauty of the area they were excavating. As I left, Jeni said:

'You should come and see for yourself.'

So, two months later, I stepped into the common room of that off-season ski hostel, high in the Austrian Alps. Peter was in front of a laptop.

'You made it!' he greeted me. 'Just finishing today's data input.'

I left him to it. Elsewhere in the common room, the team were unwinding from a day in the field. They sat in easy chairs, hair wet from showering, scrolling through their phones. Others were unwinding by working. A photographer had rigged up a makeshift studio and was recording finds under arc lights. André and Eva were discussing the parameters of pollen content. I had a chat with Matthias, whose background was statistics, about the database he was building, which involved recording 1,600 layers of soil.

Later we all walked up to the top of the valley and the Berghotel Arthurhaus. We sat at a long table and ate and drank pine-cone schnapps and talked copper – copper smelting, copper and bronze, copper hoards in the Carpathians and Denmark, fahlore copper, secondary depositions. We talked about other things – Peter touched on his work on gold in the Caucasus, and we spoke about music for a while – but we soon found our way back to copper.

We sauntered back in the moonlight. The mountains overhead were dark patches of starless sky. Our chatter rang out in the crisp air; there was singing. I was filled with a sense of camaraderie, the physical labour of archaeology and the shared motivation that informed their work: a rainbow of fascination whose arc stretched back across thirty-five centuries, linking this group to their Bronze Age ancestors, each working the same ground, crawling through the same tunnels, preoccupied with the same stuff – malleable copper and all the things it could do.

In the morning, I followed a forestry track out to Jeni's excavations. She had taken her team in the minibus. I was happy to walk. The morning air was cool and pungent and I breathed it in – pine resin, fungal spores, cow. Looking up, I tried to grasp the full majesty of the mountain, the ridge of the Hochkönig – 'high king'. With binoculars, I focused on its limestone peak. Its lifeless crackings, tumblings and dead scree were eroding fast. How much did the place itself, the presence of the Hochkönig, add to the ancient mining and processing, the 'branding' of the copper ingots?

I found the diggers in a random-looking area of scattered spruce and slanting pines – an old smelting site. The team had sliced into it, surgically removing the topsoil and constructing a poly-shelter over the wound. Walkways had been put in place around it, a network of duckboards.

'Here we have,' Jeni stretched out her arm at a mound, 'heaps of slag. And spoil over there from the dressing. The stone was crushed before heating. They were using a combination of ores.'

It looked small in the context of the forest, but the site had several furnaces.

'Five in all.' Jeni pointed to one side of the site. 'One, two, three – and over there, four, five.'

'A small industry!'

'In other parts are many more.'

The Nebra Sky Disc had introduced me to the practice of Bronze Age deposition. Mining and smelting and long-distance transportation were tasks of immense effort, and bronze objects were highly prestigious. But many were simply chucked away without ever being used, or deliberately broken before being discarded, heated up and smashed to bits with a hammer. Shiny bronze swords were snapped in two, axe-heads destroyed, meticulously crafted shields punctured with stakes.

A lot of speculation surrounds the practice – a seductive mystery whose paradox must surely reveal something profound. Votive offering perhaps, propitiation or peace rite, a performative display of wealth? All these things, probably. The anthropologist Mary

Helms suggests that it was an act of sympathetic magic, a gesture 'to support and enhance the basic life force or life energy that was thought to keep the universe properly functioning'. Metallurgy was not just tool-making. It was always an alchemy of sorts, a kind of communing with the living earth. Deposition of processed metal was, according to Helms, part of a virtuous circle: acquisition and disposal, function and exchange, transformation and re-transformation, a zero-sum game in which all things living and dead are ultimately linked, ultimately whole.

Jeni's students were working in pairs. Some had quadrats and were recording the precise positions of rocks; others were measuring, taking photos, shouting out numbers or scribbling them down on clipboards. I marvelled at their care, the forensic scrutiny of every pebble, every piece of grit, every texture or colour change in the matrix of the soil.

Five furnaces. It's estimated that it took three hundred kilograms of charcoal to produce a single kilogram of copper. One of the students told me that pollen analysis here suggested that the smelting paused regularly – and lack of wood was the assumed reason. If twenty thousand tonnes of copper was produced in the area, that would mean nearly fifty million tonnes of wood. Helms might be right about the societal meanings of smelting copper, but the wider environmental impact would also have been devastating.

I watched the archaeologists busying away. In a month or so, they'd be gone. The forest would start to reclaim the excavations and year by year, the site would revert to moss and bilberry and humus, the quiet cycle of growth and decomposition, the threads of the natural world covering its brief disruption. And I pictured another team, many centuries from now, spotting in the shape of the ground the hints of past disturbance. They'd find the post-holes for the poly-shelter, shreds of polythene, boot-marks and wear-points and trowel-dinks on the pine roots, and they might find too the faded strip of an energy-bar wrapper, a half-rotted pencil, microfibres from waterproof clothing. Beneath all that would be the furnaces and the gangue, and like these Bochum archaeologists, they would identify it as a metal-working site, where ores had been changed to copper in the long-ago Bronze Age. But what would they make of this – this twenty-first-century project, and how would they explain the motivation?

Each generation reads the past in its own way – and each is aware of what has been lost. In the Renaissance, the stature of Hermes Trismegistus was based on his antiquity. Ancient Egypt was the oldest known civilisation and Trismegistus was believed to have been there at the start. In his teaching was the *prisca theologia*, the original universal knowledge given by the gods to early humans, the golden truth that was then corrupted and diluted.

Hermes Trismegistus had always had his critics. In 1600, the philologist Isaac Casaubon finally revealed him as a bogus, concocted figure. He dissected the language of the *Corpus Hermeticum* and found it matched the Greek of the early Christian centuries. The figure of the divinely wise, pre-Mosaic sage was a sham. Casaubon's exposure of Hermes Trismegistus helped confine the ideas in his name to the shadows, along with all the deluded alchemists. As the rationalism of the Enlightenment gathered pace, no serious thinker could mention the name in public without derision.

Yet many continued in secret to hold a candle for Hermeticism – even such scientific grandees like Robert Boyle and John Locke. Francis Bacon – 'the father of empiricism' – believed that transmutation of metals was possible, but by using a more scientific methodology. While Isaac Newton was working out the laws of thermodynamics, he was equally involved in Hermetic exploring, translating the Emerald Tablet and covering pages and pages in arcane scribblings. In fact, three-quarters of Newton's entire *oeuvre* is made up of alchemical and occult writing. John Maynard Keynes, who purchased the bulk of this work in 1936, described him as:

> the last of the magicians, the last of the Babylonians and Sumerians, the last great mind which looked out on the visible and intellectual world with the same eyes as those who began to build our intellectual inheritance rather less than 10,000 years ago.

One of the key principles of the Hermetic tradition is renewal, things buried and dug up, transformed and buried again – constant process, constant change, the cycle of energy identified by Helms. As a set of universal ideas, Hermeticism itself has undergone the same pattern – renewed, hidden away, then revived in slightly different guise – from ancient Egypt to early Christian Alexandria to Renaissance Europe. Who's to say that it is not ready for a twenty-first century reinvention?

In 2003, the phrase 'Green Hermeticism' emerged among a group of ecologists who were struggling to find a language and a set of principles to articulate the environmental reset, to formulate a 'spiritual ecology'. It coincided with the final acceptance of alchemy as a legitimate historical enquiry. According to Christopher Bamford, one of the movement's initial advocates, they were all drawn to the Hermetic idea because it saw 'no boundary between the cosmos, humanity, and earth'. It placed emphasis on 'one reality: a single, whole, a uni-verse'. It also transcended global boundaries – there were lively traditions of alchemy in Christian, Muslim and Taoist cultures.

Bamford identified an age-old distinction. Some philosophers and scientists are concerned with 'substance' while others concentrate on 'pattern or process'. He spoke of a line of thought, of '"pattern thinkers" from the Gnostics and alchemists through the Renaissance Hermeticists, and on up to Lamarck, Blake, and Romantic Scientists like Goethe and Novalis. This was the clue that put [him] onto the alchemists, who . . . had an epistemology that consciously sought to give expression to nature's own way of doing things'. There was a period, he believes, between 1780 and 1820, when a battle took place for the soul of the Enlightenment, which meant that one side won and the other – the 'pattern or process' thinking – was subjugated.

One figure whose career more or less spans those decades was William Blake (he was still working until his death in 1827). A consistent theme in his writing and his images was rage against the errors he saw rising around him. He reserved his blame in particular for the triumvirate of Newton, Locke and Bacon. Their sin, he believed, was to exalt reason over imagination, reducing the complexities of the universe to numbers, its movements to formulae, the strangeness of life to causality. In the book *Jerusalem: The Emanation of the Great Albion*, he pointed to dead metals and machines as the result:

> For Bacon & Newton sheathd in dismal steel, their terrors hang
> Like iron scourges over Albion, Reasonings like vast Serpents
> Infold around my limbs, bruising my minute articulations
> I turn my eyes to the Schools & Universities of Europe
> And there behold the Loom of Locke whose Woof rages dire
> Washd by the Water-wheels of Newton.

Blake was well acquainted with Hermetic views. Elsewhere in *Jerusalem*, he refers to the Emerald Tablet (unaware of Newton's own fascination with it). The figurative beliefs of alchemy – spiritual transformation, creation and renewal – chimed with his own. One of the mythical beings he invented for his poetic pantheon was Los: a metalworker toiling in the heat of his furnace, an alchemist drawing on cosmic forces to change one thing to another, and an iconoclast, smashing artefacts to make something new.

I pictured Los at large, here in this ancient Alpine smelting site. Nature's materials were being altered, mined ores smashed to gravel before being thrown into the furnace to be transformed:

> Los with his mace of iron
> Walks round: loud his threats, loud his blows fall
> On the rocky Spectres, as the Potter breaks the potsherds.

Los is the master of fire and he also represents the poetic imagination. He destroys but, like philosophical Mercury, he also oversees change and creation. Blake is urging us to find beauty and strength in the physical world, to rediscover in Los's works what we need for our own time:

> All things acted on Earth are seen in the bright Sculptures of
> Los's Halls & every Age renews its powers from these Works.

Copper was Blake's medium. He was known in London as a jobbing engraver of copper plates, a man who spoke about strange things and wrote impenetrable verse but whose prints were nonetheless exemplary pieces of craft. He did short runs of his own work but supported himself and his wife Catherine through commissions. He spent as much time preparing, engraving, etching and printing from copper plates as anything else in his life. By his own account, he 'never suspended his Labours on Copper for a single day'.

The plates he used were thin, no thicker than a half-crown coin. When he took delivery of them, the first job was to polish them with a grinding stone, then with pumice, then with a smoother stone and charcoal before again polishing with steel, and finally with a piece of stale bread. The prepared plate looked like a mirror, and as Blake leaned over it to make the first cut with a burin or needle, he would see his knuckles reflected in the copper, a colour like nothing else on earth.

He was involved with every stage of plate-making and printing. He owned a star-wheel press, which he leaned into and turned himself, exerting the necessary force to push the ink onto the paper. The press followed him around his various lodgings in London, and for a two-year stay in a cottage in Felpham, Sussex, and then finally to a few modest rooms he took on the Strand. Drying prints hung from lines around the walls. Inks stained his fingers. His days were filled with the careful application of acids to the plates, and all around him was the reek of nut oil and the metallic smell that came from the copper as he heated it for etching and printing.

Blake was a man who lived in the celestial realm of his own imagination, but he saw no boundary between that world and the physical world. Like the alchemists, he believed strongly that art and craft were one – 'mechanical Excellence is the Only Vehicle of Genius'. He understood the calamitous falsity of isolating opposites – a tendency that he sensed growing all around him. It angered him to see Cartesian dualism in the ascendant: soul separate from body, spirit from matter, humans from nature. He wasn't to know then the full impact of that division – the licence it created to scale up extraction and consumption of the earth's resources. But he knew it was fundamentally wrong, an affront to the joined-up qualities of creation.

His own thoughts and his actions were of a piece: the engraved line in the copper and its message were both alive in their own way.

Each he saw as playing its part in his version of the Great Work, the task of trying to restore cosmic unity. The chemicals etching away the plates had a metaphysical parallel:

> The notion that man has a body distinct from his soul is to be expunged; this I shall do by printing in the infernal method, by corrosives, which in Hell are salutary and medicinal, melting apparent surfaces away, and displaying the infinite which was hid.

The 'infernal method' of printing was one which Blake himself invented, a feat which is in its field as innovative as his poetic works. At the time, illustrated books were produced in two distinct parts: text and images separately. Blake wanted to create something in print that was more organic, more akin to a medieval illuminated manuscript. The 'corrosives', the acids, he used in a new way. Rather than cut into the metal to produce the design, they cut into the rest to leave the design raised, in relief. In this way more copper was dissolved and the acid-resistant 'ground' was left for longer, generating more toxic fumes. But it enabled him to combine images and words on the same plate, to print it in one, 'to display the infinite'.

William Blake was in his mid-sixties when he began work on his *Illustrations to the Book of Job*, recording the day as 2 May 1823. They were completed nearly three years later. He did not use relief etchings for this series, his 'infernal method', but intaglio engraving. Sitting at a table in his bedroom near the Strand, with a window above it, the light flooding in, he could see the River Thames shining 'like a bar of gold'.

The *Job* engravings are now seen as one of Blake's greatest achievements, the most complete expression of his ideas, the finest of his engravings. Ruskin considered that in these works, with the use of 'glaring and flickering light, Blake is greater than Rembrandt'. Writing in the early twentieth century, Joseph Wicksteed reflected on their creation and on Blake himself, concluding that it would be hard 'to find a more beautiful chapter amongst the annals of art than the evening of his life spent in engraving these Job designs'.

Blake laid the polished plate on a small leather sand-sack, which he could swivel and turn as he worked. He had already sketched out the designs on fine paper, pressed them down on the copper plates, then pricked out the outlines with drypoint. With the mushroom-shaped handle of the burin against his palm, he pressed the blade into the plate. A range of burins were used for engraving, each with a subtly different profile to its cutting edge – the square and lozenge, the spit-sticker and the scorper. Decades of practice lay behind the minute adjustments Blake made – to the position of his shoulders, to the pressure he applied according to the resistance of the metal. It was the perfect combination of body, metal and imagination. Burrs of copper glittered around the plate, a reddish-gold flotsam of shards so sharp they could draw blood.

In his etching and engraving, Blake saw the task as using the physical world to reveal the hidden one behind it; through the copper plates he could display his view of that arcane and wonderful territory. Like an alchemist, he was constantly experimenting, perfecting the processes of transformation – and he was secretive: he might speak or write of his techniques but he never let anyone see him practise his relief etching.

Blake believed in the material vocation of his art, the metaphysics of his print-making methods. But of all the hundreds and hundreds of copper plates he worked on, only a few have survived – and of his 'infernal method' only a fragment of a single plate. An entire industry of analysis has grown up trying to recreate his particular craft. Blake scholarship is an archaeology of sorts.

On the wall of my studio in Cornwall hang two small prints from William Blake's *Illustrations to the Book of Job*. They are not original but facsimiles from the Trianon Press. They belonged to my father-in-law, Anthony Hobson, who was a collector of art and rare books and also a world expert on the libraries of Renaissance humanists. His particular specialty was tracing the passage of early modern thought across Europe through the signature style of Italian book-bindings. When he died, the Blake prints were found among his papers. Knowing of my interest in Blake, Charlotte had them framed up and gave them to me for Christmas – but what I would give now for just an hour to ask Anthony about the prints and what he made of them, and about the humanists and the Hermetic tradition.

Anthony was a man I admired greatly, whose company and knowledge I always enjoyed. We spoke often about his travels, the books he loved and the friendships he'd built up over a lifetime – but not so much about his work. Into his nineties, he would spend much of the day in his upstairs study. What went on there was an endeavour shared by a small community of international biblio-philes and scholars.

Outside the study, on the landing, were overspill bookshelves and they themselves spilled over, down the corridor. When I first visited him, before Charlotte and I were married, the corridor shelves were new and mainly empty. Each time we visited, there were additions – published books, books about books, sale cata-logues and journals – arranged by subject. I would sometimes flick through them, marvelling at the bibliographic labyrinth they explored. The pages were interleaved with dozens and dozens of additional notes and corrections, in Anthony's own hand.

The years passed and the shelves filled. I would pass them in the middle of the night, padding downstairs to heat up a bottle of baby's milk, then sitting on the high-backed chair in the corridor, reading the spines, watching the head in the crook of my elbow drink and the eyes close and the world return to its axis. In later visits, the children would run up and down the corridor, riding an old dog on wheels, pushing toy cars along the floor, curling up in the bottom shelves to hide. 'Sh-sh!' we'd hiss, if their games took them across the landing. 'Grandpa's working.'

By the time Anthony died, the bookshelves were full. The collection went to the Bodleian Library. The Blake prints turned up in an envelope and without explanation. But when I looked into the Trianon Press, I understood at once how their meticulous facsimiles would have appealed to him.

Set up in Paris by the William Blake Trust in 1947, the press was run by one-time spy Arnold Fawcus, an eccentric obsessive and investor in vintage sports cars, semi-ruined chateaux and blue flowers – rare, blue delphiniums. It often took his craftsmen a month to produce a single, acceptable proof from a plate, while the printing was done using repeated inkings – some of the blacks required forty layers. Blake himself would have approved. The Trianon prints were physical devotions, as his own were. Each set was dedicated to conveying the abstract world of ideas and beliefs through engraved copper. Trianon's prints were so true to the original that a small watermark had to be inserted to identify them as copies.

When I stand from my desk to look at the prints, I use a magnifying glass. Each image is no bigger than the cover of a hardback book. Moving my head back and forth, I wait until the image, or part of it, bursts into focus. Such scrutiny reveals what a technical feat they were, the tiny strokes, the minute stippling. Blake himself used an eyepiece to work on the plates; such an optic allows an intimacy that the naked eye does not. It breaks up the image into regions, over which you can hover, shifting from one to the next, looking down on them like a bird, so that – like the moldavite on my windowsill – these diminutive oblongs of ink, the tiny scratches of light and shade, suddenly take on the scale of landscape.

Throughout his life, Blake returned again and again to the figure of Job. In his twenties, he produced several ink sketches of him, an engraving a few years later, a tempera painting and then two sets of watercolours. He had already done nearly fifty representations of the poor patriarch when he turned to the allegory for the last time, to tell the whole narrative (or his version of it), his own swansong series of twenty-two engraved copper plates.

To Blake, Job is a man in error – not just in error, but in ignorance of his error. He has applied all his energies to the outer world, to possessions, to the letter of the scriptures and the law. He has neglected the inner world, the world of spirit and imagination. For this he is exposed, and for this he suffers.

Blake's words and images possess, like the greatest creations, an inbuilt ambiguity, an innate capacity to hold multiple interpretations. Hermeticism and the mystics display something similar – the power to multiply their meanings until they either confound or provide a glimpse of revelatory truth. Such work also has the ability to apply to the particular. Hold it up and it will reflect back whatever is concerning you. As Goethe's *Faust* has its contemporary resonance, and Hermeticism its new green followers, so it's easy to look at Blake's Job and see in him the collective crisis facing the planet.

The story begins in prosperity and contentment. Blake's Job is industrial man, convinced by the empirical truths of the Enlightenment. In Plate 1, he is surrounded by family and wealth, and the books open in his and his wife's laps show they are obedient to rules and conventions. But beneath is a clue to his erring: *The Letter Killeth / The Spirit giveth Life*. One of the two images that I have is Plate 2 – in it his comfortable life is still intact. He still has all that gives him happiness, the book remains open and God looks down on him from on high (albeit with a troubled expression). Blake has written another caption from the biblical Job: *When the Almighty was yet with me, When my Children were about me*. But now there is a new figure that dominates the scene between them – sweeping into view in clouds of flame, Satan, making his bid for Job's future. The more I look at this image, the more it appears to reflect back our current position: aware of the coming calamity but unable quite to relinquish our ways.

Flick forward, through the next plates – past the disasters, the destruction, Job's loss of his children, his journey into despair and isolation. Each of those images is dark and night-like, the ink on the copper thickly spread. The second image in my studio is Plate 14. It comes in the series like a sudden flood of light. God is not at the top this time, but in the middle. The corporeal Job is below him, sitting with his wife and three friends. They are in a cavern and have nothing with them but their meagre clothes. Their hands are raised in supplication and their gaze is directed upwards – not so much to God but to the drama that takes place above him. Along the top section runs a row of seraphim. Behind them is dawn's pale rising: *When the morning Stars sang together, & all the Sons of God shouted for joy*.

Few other figures in Blake's entire work are so rapt, so celebratory, as those seraphim. The row of them is cut off by the edges, giving the impression that they are limitless in number. Arms upstretched, open-faced, they are the very opposite of the awkward pose in another of Blake's images, the hunched-over figure in *Newton*.

In Plate 14, Blake is giving us Creation. The margins are filled with images from the first chapter of the Book of Genesis: *Let there Be Light . . . Let the Waters bring forth abundantly . . . Let the Earth bring forth Cattle & Creeping thing & Beast.* This is not the six-day Judaeo-Christian myth of origin, God's six-day assembling of the material world, nor is it the original dividing of darkness and light, heaven from earth, land from water. It is everything, always.

To Blake, the universe is in a state of constant creation. That is the revelation of the physical world. Every moment, it is bursting into existence, growing, being born – and death and decay are part of the cycle. Opposites create. Nothing is still.

This is not the biblical Job. Blake's version presents Job's mistake as neglect of the imagination, the little piece of God in us all. It is, according to Blake, the thing which makes us human, which enables us to play our part in the perpetual process of transformation. We are all the sons of God and we are the morning stars too, and we are singing together. The image on Plate 14 is a manifesto for kinship with people and all living things. It is also a manifesto for the ecstatic state that flashes before us when the clouds of this world part and for a moment we glimpse the unity beyond: joy is joy because it is also truth.

Joseph Wicksteed was one of the first to give the images of the Job series a full critical interpretation. He trawled Blake's work for precedence of each element. He noticed, for instance, that place-ment of the right and left limbs had been used by Blake throughout to convey, respectively, the spiritual and corporeal. Wicksteed's entire study quivers with the thrill of his own revelation, the excite-ment that comes to anyone looking openly at Blake's work, just as Blake himself urges us to look at the world.

Of Plate 14, he writes: 'Blake touches the height of his genius, and, indeed, of genius itself.' The combination of the morning stars and the seraphim is enough to 'waken harmonies of so intense a joy, that it seems to tremble on the verge of a great sob, in which the eternal grief of the universe is merged in its eternal rapture'.

Early the next morning, I joined Peter and his group to go to work in the Mitterberg copper mine. We took one of the minibuses and drove down to Bischofshofen to pick up a three-metre steel beam they'd commissioned to shore up the mine's ceiling. The foundry would only take cash, so we all piled back into the van to drive to an ATM. When we left town, the beam lay on the floor, under our boots.

Peter was driving. Fabian's speaker was on the dashboard, belting out high-intensity music. I liked it that these mining archaeologists chose to listen not just to hard rock, but to metal in its various forms – heavy metal, thrash metal, doom metal. Though not all of them, not Anna, sitting beside me. 'Some of us have – how can you say? – different tastes. The one who everyone agrees on is Johnny Cash.'

Peter was swinging the van up through the hairpins. The road was rough and unmade. He was tapping the steering wheel with his palm each time he re-entered a straight. The slope steepened; below us the settlements of the Salzach valley grew smaller. On the sharper corners, there was a swish of gravel and the beam shifted beneath our feet. It made me think of the final scene in *The Italian Job*.

Then another song started – with a tinkling organ and falsetto vocals.

> *Unter den Gräsern*
> *Nähre ich mich von Dunkelheit*
> ('Among the grasses / I feed on darkness')

234

This was suddenly much gentler, more melodic. The van was filled with the song, but also with singing. I realised now that everyone was singing – Peter, Fabian, all the metal fans and the others too. And not just mouthing the words, but singing with gusto, harmonising the chorus.

'It's traditional,' Anna explained. 'We put it on every time we go to the mine.' The song was '*Weg nach unten*' ('The Way Down') by the mock-metal band Knorkator – whose appearance for a Eurovision Song Contest qualifying show famously prompted a *Bild* headline: '*Wer ließ diese Irren ins Fernsehen?*' ('Who let these lunatics on TV?')

Anna resumed singing with the others:

> *Kleine Welt*
> *Die ich ertasten kann*
> *Ab und zu*
> *Find ich Kupfererz*
> *Diamanten und Granit*
> ('Little world / That I can touch / Now and then /
> I find copper ore / Diamonds and granite')

The song's last verse was something about finding a kangaroo and realising that maybe they'd dug too far. The song was like a pre-match ritual and when it was over, it left the sense of a team vibe in the van. Peter drove a little more slowly.

We pulled up below a squarish pavilion with tall windows. The *Wasserschloß* was built in the nineteenth century over the entrance to an underground canal. It was now a base for excavation of the nearby mine, and as soon as we entered it, everyone got into character. From pegs, they pulled down mud-soiled boiler suits. Bina bent his head to wrap dreadlocks in a pink-and-white scarf; he did up his orange overalls and on the back was a wolf's head and the legend *GRUBENHUND* ('mine-dog'). Peter tied a keffiyeh over his head and changed into a piratical outfit of grey tunic and breeches. He held up a brass lamp. 'An old miner came to the institute in Bochum and said, "I don't need this any more. You guys have it." So we always take it underground – for luck.'

Max was a big guy and he did not change his clothes. 'I am not built for underground,' he told me, without regret. He set up a camera to record each one of a heap of rocks taken from the mine. His denim jacket read *Heavy Metal Tyranny*.

The minibus went ahead to the mine with the steel beam. The rest of us followed on foot, spread out across the road like a 1980s glam-metal band. The Bronze Age copper works was accessed via the nineteenth-century entrance, which cut into the hillside. Above it was a castellated stone lintel with the name *Arthurstollen* below a cross-hammer emblem. We transferred the steel beam onto an old hand-wagon and pushed it on rusty tracks into the darkness. The bare rock brought our voices close and amplified the crunch and rattle of the wheels. Ten minutes in, and a squeeze dropped to one side. This was where the mining had broken through to the Bronze Age galleries.

Setting the steel in place involved dropping it down the winze, lowering it with a rope, while at various points below, the team guided it on. There was a lot of shouting and the clang of steel against stone. In the Bronze Age, they had no access to machine-tooled steel beams, and no hand-wagons on rails. They jammed wood into the rock to prop up weakened ceilings, but with the same team effort. Leaving the others to finish the job, Peter took me off to see the rest of the Bronze Age workings. With us came Phyl, a PhD student from Innsbruck.

In his swaggering costume, Peter shimmied down a small shaft, still talking as his head disappeared. We followed him down into a warren of tunnels. He had a way of moving underground that was deceptively efficient. He was not hurried but he was fast, his long limbs appearing to predict the contours of the coming space, like a snake. He knew every inch of this mine. He'd been studying it for over ten years.

'So the chalcopyrite lode goes this way . . . but here, you see? Here is a fault and the ore stops. But they knew. In the Bronze Age, they knew it would start up again somewhere. So you have exploratory galleries here and here.'

Mines from this period are very rare. It's in the nature of extraction that each age's work is destroyed by the next. Stopping to rest, we sat with our backs to the wall and drank water and snacked. I asked Phyl about her work.

'I've drawn up a list of sites that maybe have Bronze Age mines. I am working through them, but rather slowly.'

'Archaeology seems a slow business,' I said. 'Everyone here is so methodical.'

Peter smiled. 'Some things can't be rushed.'

Phyl scraped her boots on the gravel. 'This is the only actual mine I've been in. It's impressive. Maybe a bit cramped.'

She loved everything to do with mountains – inside and out, and she was a keen climber. But recently all her time had been spent working through her list, examining the topography and geology. 'I have been to so many places. I found some possible ones but nothing certain. I am still looking. Sometimes I think this work is only ever looking!'

In Cornwall, at just such an underground resting, I'd first heard tell of the 'blue mine'. We were exploring some abandoned works near

Helston when someone mentioned a copper mine where second-ary deposition had left extensive areas entirely blue. The location was a secret, to protect it. I found in the following days that the image wouldn't go away and the longer it sat in my mind, the more it made me think of *die blaue Blume*, 'the blue flower'.

To German Romantics of the late eighteenth century, *die blaue Blume* was a powerful motif, a much-cited symbol of the earthly search for rarity and wonder, love, harmony with nature, truth – pretty much everything good, in fact. The flower part of it was perhaps less potent than the blue – the colour of the infinite, the unattainable, the distant sea and sky. The blue flower has been used by many authors since, from Walter Benjamin to John le Carré, Penelope Fitzgerald to C. S. Lewis.

First mention was by the eighteenth-century German writer Novalis. In his novel *Heinrich von Ofterdingen*, young Heinrich falls asleep and has a dream in which he travels far. He dies and is reborn and is enraptured by beautiful landscape. He is made destitute, fights in a war, is captured, loves intensely, loses his love – 'his life becomes an unending tissue of the brightest colours'. At last he reaches a spring and sits on the soft turf and watches the water rising like a fountain into the sky. There, beside the spring, he spots a small blue flower. Of all the things he has seen and experienced, it is this blue flower that he longs to keep, and he reaches out to pick it – at which point he is woken by his mother. From then on, he sets off to explore the world, on the hunt for something he can't quite identify: 'I yearn to get a glimpse of the blue flower. It is perpetually in my mind, and I can write and think of nothing else.'

At one stage, Heinrich comes across an old miner from mineral-rich Bohemia, a fellow searcher whose blue flower happens to be rocks:

From his youth he had been very curious to know what might be hidden in the mountains, from where water poured its visible springs and where gold, silver and precious stones were found.

The miner says that his own search for minerals has given him a happiness that no other-worldly pursuit could have offered. He is 'satisfied with knowing where metallic riches are found and bringing them to the light'. But once found, metals are toxic: 'in the form of property, they become a terrible poison'. The two men then take a night-time trek into the mountains to find a cave. Deep in the cave they come across a hermit, who radiates a sage and contemplative joy. The hermit has many books. Flicking through one of them, the young Heinrich turns the page and is astonished to find an image of himself.

Novalis's novel is pure quest. It's an almost parodic distillation of romantic searching, written as a series of stories within stories, a dense and intricate structure underpinned by a pervasive yearning and the paradox of all our earthly endeavours – that what we strive for is always less fulfilling than the striving itself. Appropriately, Novalis never finished the novel.

I did manage to reach the blue mine. A friend rang out of the blue, as it were, and said they were going down that afternoon. The entrance was hidden away, on private land. At a depth of about thirty metres, we reached a passage where the walls began to show streaks of colour. The streaks thickened. I'd seen patches like this before in other old copper mines – but never on this scale. In geological terms, it's caused by water seeping through the rock and leaching out the copper as copper sulphate, a semi-formed chalcanthite (from the Greek 'chalkos', copper, and 'anthos', flower). Spread over the walls, it created an expansive, blue-sky effect all the more powerful for being underground. There were multiple shades of blue – light powder blue, deep bottle blue, International Klein blue, marine blue, celestial blue. It tumbled down the rock in a shiny blue cascade of little wave-like ridges, and from the ceiling it hung in blue stalactites and curtains of blue which draped the rock, falling and folding in cloth-like undulations of blue. On flattish pieces of rock were tiny blue pools of clear water and in them clusters of blue sulphate eggs. Walking through that blue gallery was dream-like until, looking up to inspect the ceiling with its constellations of shining blue, a drop of water fell – splat! – into my eye.

Surfers speak of a place known as the 'green room'. If you're very good and the wave is right, you can tuck in under the lip and find yourself in hollow space. Your board is pointed at a narrowing opening ahead but otherwise you are surrounded by water, enclosed for a few seconds in a bubble of translucent blue-green. Then it's gone. It's not a place I've been in, nor am I ever likely to. But if surfing has a questing element to it – the obsessive pursuit of moments of fleeting perfection – the green room is its blue flower.

Over the millennia, the copper mine at Mitterberg has drawn in its fair share of searching souls. In two separate phases – using antlers, flints and fire – Bronze Age miners put in the effort here to obtain ore to make copper to make bronze and, in many cases, then to throw it away. The mine lay abandoned for over three thousand years, until industrial-era workers brought hammers and explosives and opened up the lodes, and exposed the earlier working. Then came the Bochum archaeologists with their cameras and computers, their laser levels, 3D imaging and carbon-14 dating. They came to conduct something of a Hermetic experiment, a re-mining and a re-searching, a meticulous piecing together of ancient knowledge.

Peter, Phyl and I were now deep inside the original mine. It was all very tight. We reached a small gallery and a narrow passage led down from it and we put our hands on the walls and lowered ourselves down one by one. Every metre meant a contorted effort. A little way down, it turned a corner, narrowed further. It took quite a time to get to the last section, which fell steeply to a stop. Each of us was arranged above the other, propped against the sides of that slim rocky tube. I was in the middle. It was the only time during all my mine trips that I felt any claustrophobia – something about the tapering of the tunnel and dead end. Peter was curled up at the bottom, reclined and at ease.

'This was all backfilled when we found it,' he explained. 'Emptying it was a little difficult. Two of us would do it, passing up buckets. If we did twenty-four buckets in a day, we were doing well. It took three summer seasons to reach the end.'

Three seasons! It struck me again: the work of these Bochum archaeologists was just as laborious as that of the Bronze Age copper workers. 'What did you find?' I asked.

'Not much. A broken measuring stick and wooden compass. It looked like they'd been broken deliberately. Maybe it was because this digging yielded nothing. Like a superstition about the tools.'

We uncoiled ourselves, retraced our route. To re-enter the modern mine, we had to slide down a smooth section, then drop blind to the floor below, like a laundry chute. When Phyl did it, she slipped and landed askew. We knew at once she was hurt.

'Phyl?' Peter was at her side.

'It's ... OK ...'

But it wasn't. Her face was twisted in pain. She was gripping her leg. 'I think ... it's here.' She grimaced.

'Don't move.' Peter went off and came back with a couple of others. We carried her carefully to the wagon, put coats under her leg to cushion it. 'Get her to hospital,' said Peter. Three of them pushed her out. We watched them go, the lights flickering and growing more distant on the adit walls.

Back at the hostel, I was getting ready to leave when the minibus came back with Phyl. We all went out and found her on crutches. Her leg was in plaster.

'The tibia's fractured.' She eased herself down onto a bench. 'Actually it feels not too bad.' She was smiling. 'Maybe it is the pain-killers!'

I said goodbye to her, and to Peter and Jeni and everyone else. As I left, Phyl was telling a story about one of the nurses. Everyone was laughing.

At the bus stop above the hostel, I looked up at the Hochkönig. The sun had just dropped behind it and its rays fanned out from the spikes of the ridge – and it made me think of Blake's seraphim with their upstretched arms: *When the morning stars sang together . . .*

William Blake was no worldly seeker. He did not, like Paracelsus, thirst to go off exploring the world. He was a walker, but never went further than the fields and villages to the north of London, or the downs to the south. He did not, as Goethe had done, look for the secrets of the cosmos in the patterns of plants and rocks. Truth came to him unbidden, out of the ether, in the form of visions. His calling was to communicate what he saw, using verse and painting and copper plates, using aphorism and metaphor, animating in his long verses a whole series of symbolic figures.

Few listened. By the time he started work on *Illustrations to the*

Book of Job in 1823, Blake had become an obscure figure. His old friend Charles Lamb wondered 'if he be still living'. Blake had for some time realised he would never see his work widely appreciated. Instead he hoped that it would be 'a Memento in time to come & to speak to future generations by a Sublime Allegory'.

His health was in decline. His handwriting from this period looks shaky; his few visitors would often find him ill in bed. He was short of money and had recently been forced to sell his cherished collection of old prints, though he kept his favourite – Dürer's moody head-jumble *Melencolia I*. He was looked after by Catherine, his wife, and their devotion to each other never wavered.

Several hundred sets of *Illustrations to the Book of Job* were printed. Some had been pre-sold to those loyal followers who had stuck by Blake. But they did not prove popular. The style was outmoded. It was 'too much Finished, or over Labour'd'. There were one or two positive responses, but a year or so later, no more than thirty had been sold; some of these had been returned by dissatisfied customers.

What makes each age blind to its own present? What makes ours so complacent? A Jobian hubris, a fear of change, the sort of disruption Blake was advocating? Each time I cross my studio floor to look at the two prints, I see something new, and am struck afresh by the twin notes of doom and joy.

After he completed the *Job* plates in 1826, Blake's work rate did not let up. He was using watercolours for a series illustrating Dante, finessing a large oil of the Last Judgement. Among his final pieces was a visiting card 'in which the images of children, and of spirits holding the threads of life, act as an allegory for human existence'. He had always loved children. He and Catherine had had none.

He was afflicted now with a series of complaints that had dogged him on and off for years. He lacked an overall diagnosis. As he grew more frail, his mind remained sharp. At one point, he took a sudden downward turn, but recovered. In a letter, he reported that he 'had been very near the gates of death' and remained 'very weak and an old man, but not in Spirit & Life, not in the Real Man, The Imagination which Liveth for Ever'.

In his final weeks, he seems to have been happy. He was 'going to the country he had all his life wished to see'. Catherine spent her time at his bedside.

'You have ever been an angel to me,' he told her. 'I will draw you!' And he took his pencil for a final portrait.

His first biographer, Alexander Gilchrist, gives an account of his final hours. The twelfth of August was a Sunday. He and Catherine were in his room. Beyond the window flowed the Thames, golden in its sunlit strip, and Blake's face suddenly relaxed. It became 'fair', as if some great struggle had ended. He appeared to be in a state of joy, and like the morning stars in Plate 14 of the *Job* series, he began singing. They were not familiar songs, but hymns devised on the spot, hymns of wonder.

'My beloved!' he said to Catherine, as if to explain not just the songs but the force that had driven his life's work: 'They are *not mine. No!* they are *not mine.*'

He died that evening.

A good deal of effort has been made in trying to identify the cause of Blake's long-term illness. His own letters and first-hand accounts mention various symptoms. He clearly suffered from liver disease. Reports of yellow skin towards the end of his life suggest jaundice and blockage of the biliary tract.

Copper had been his daily companion, the medium of his life's work, his images and his words. Sometimes, as with the *Job* series, he had cut straight into the plate, producing metallic shards and curlings. His earlier innovation of relief etching – the 'infernal method' – required nitric acid applied to the copper over sustained periods. In badly ventilated rooms, the fumes were the smell of work. The likely cause of Blake's death, according to a group of literary and medical experts, was liver failure caused by sclerosing cholangitis, which itself developed from a long build-up of hepatic copper.

10

GOLD

To the Incas, it was the sweat of the sun god, to the Aztecs his excrement. In the sun, gold cannot be created – silver and lead can, but not gold. Only in the most extreme events – core-collapse supernovae, the collisions of mega-stars – do clusters of free neutrons occur with enough density to make gold. Once formed it hurtles out through space in clouds of dust, flashing in the emptiness, a shower spreading into the eternal black. Most of the gold that struck the earth in earlier epochs made its way down into the core, where it remains. Only a small proportion has stayed on the surface.

So hard to form, gold is almost impossible to destroy. It does not oxidise. It does not tarnish or corrode. It might be separated from its matrix, washed down into rivers, deposited and pressed into shales, and those shales may be twisted and shifted or transformed by heat, and they may be scraped away by glaciers or scoured by floods, but the gold will continue to gleam in its granular form, or fuse into honey-coloured nuggets. Scarce though it is, gold is absent from few regions of the earth.

Glittering in shallow streams, shining from the mud of alluvial silts, gold is a small parcel of sunlight. It cast its spell on people, stirring curiosity first, then greed. Something about it made these acquisitive, two-legged creatures reach out and pick it up. And those who did found it strange stuff. Oddly heavy and oddly soft. You can shape gold like no other metal. An ounce of it can be flattened to a sheet ten feet square, one thousand times thinner than newspaper; if pulled out, it can stretch to a filigree thread fifty miles long.

Another strange aspect of gold is that it turns up late in the archaeological record. Copper and meteoric iron, which are much harder to work, are found at burial sites in the Mesolithic, but not gold. When it does appear, in later burials, it is on an elaborate scale. One of the earliest known examples of fashioned gold comes from the six-thousand-year-old tombs at Varna in Bulgaria: no evidence of gold at all in Europe, and suddenly here are more than three kilograms of worked objects in a large necropolis. A number of the bodies were wrapped in shrouds appliquéd with gold, with mementoes of gold and copper around them. In grave number 43, more gold was found than in the rest of the world combined for this period. Lying on his back, the skeleton is holding a gold sceptre with beads around his bones, three necklaces, pendants and discs – and an impressive gold penis sheath.

Gold enters European history in the role it has played ever since – display and social distinction, demonstrable assertion of material wealth. From Neolithic burials to Rolex watches, gold exalts those who have it. 'Gold is the source of crime, the plague of life, the ruin of all things,' wrote Phocylides in the sixth century BCE. 'Would that you were not such an attractive scourge, because of you arise robberies, homicides, warfare, brothers are maddened against brothers.' But gold is also financial safety and security, the currency of last resort. The king of metals, it presides over human life by channelling want – not for an object or a person, a time or a place, but for the possibility of all of them.

In South and Central America, gold was more about reverence than desire. It was seen as transcendent, the mystical in physical form. Elaborate gold regalia, which were a feature of many pre-Columbian civilisations – head-dresses, collars, tassels and bells – are understood by ethnographers to link the state with the divine. In the Coricancha complex in Cusco was a sacred garden dedicated to the sun god Inti. There were flowers and fields and life-sized replicas of llamas and insects and other animals, with everything in it made from gold and silver. Then the Conquistadores came and took the sacred gold and melted it down and sailed it home in ingots. Such was the scale of pillage that by 1560, gold brought back from the New World had doubled Europe's entire supply. In Cusco's sacred garden all that remained were a few gold wheat stalks.

In February 1869, a Cornish miner in Australia was digging when his pick struck a rock, and the handle broke. John Deason, born in the Isles of Scilly, fetched a crowbar and prised out the stone, which now shone and gleamed in the sun. The Welcome Stranger, as it became known, weighed seventy-two kilograms and Deason sold it to the London Chartered Bank of Australia for £9,553. It was the largest gold nugget ever discovered.

Nuggets are for the lucky few. For the rest, gold prospecting is an arduous and usually fruitless pursuit. It can also be dangerous. Herodotus wrote of the desert region to the east of India famous for its gold-rich sand. Retrieving the gold was difficult on account of giant ants – vicious beasts a little larger than a fox and extremely quick on their feet. Prospectors rode in on camels. They gathered the gold in the hottest part of the day when the giant ants were asleep. Once the intruders were detected, the ants woke and gave chase. The male camels were too slow and were taken by the ants, but the gold-hunters made sure to be riding the female camels and to load them with the gold. Females who had recently given birth

were the quickest because nothing could stop those camels returning to their young.

The largest gold mines in the Roman Empire were at Las Médulas
in Spain. Hundreds of miners burrowed and chipped their way into
a mountainside, creating a network of narrow shafts. 'For months,'
reported Pliny, 'the miners cannot see the sunlight.' In pursuit of
the sun's glow on earth, men were hidden from it. Then the entire
mountain collapsed. Another, on Thasos, did the same, and Herodotus reported that that also was a result of gold-mining. Pliny
called the method of mining at Las Médulas *ruina montium*, 'wreck-
ing of the mountains'. Once the hillside was reduced to rubble, 'the
miners gaze as conquerors on the collapse of nature'.

'Gold is immortality,' states the Vedic text *Shatapatha Brahmana*,
which also offers details and meaning for its ritual use: a Brahmin
hangs a disc of plate gold from his neck, and the disc is the sun, and
the seed of Agni, and it is truth and light and eternal life, perfection
of both form and substance. In early India, as in China, alchemy
was focused on the pursuit of 'potable gold', an elixir that would
postpone death indefinitely for those who drank it.

In Middle Kingdom Egypt, the display of gold was reserved for
kings, but the Queen of Sheba had enough gold of her own to present Solomon with 'a hundred and twenty talents of gold' (estimated
at over four tonnes). Today, 78% of the global gold market is jewellery, and only 10% is used industrially. Of that, a tiny amount, fifty
cents' worth, is contained in every mobile phone. Less than two
hundred thousand tonnes of gold has so far been taken from the
earth. If put together, it would form a cube with sides of only about
twenty-eight metres.

South Africa, Australia and China are the biggest producers
now, but in the Middle Ages the rivers of West Africa yielded a great
deal of gold. Berber and Tuareg camel-trains burdened with gold

bars crossed the Sahara to Marrakech, Fez and Tunis. When Mansa Musa, the fourteenth-century ruler of Mali, left for the haj, he took with him twelve thousand slaves, each slave carrying a block of gold, and the panniers of his camels were filled with gold dust. On his arrival in Cairo, he began to spend and his spending power crashed the Egyptian economy. Mansa Musa appears in the Catalan Atlas of 1375, pictured in his kingdom holding a nugget of gold.

It was at this time that appetites in Europe were beginning to grow – for trade, for new territories, for the range of goods that could be brought back from overseas. Portuguese took slaves south to swap them for gold. Later, the trade was reversed: gold was exchanged for human souls. The Gold Coast became the Slave Coast.

At the beginning of the eighteenth century, a group of warring West African kingdoms came together to form the Asante Empire, achieving unity by promising allegiance to a golden stool. In 1817, a visitor to the Asante court recorded that:

> The royal stool, entirely cased in gold, was displayed under a splendid umbrella, with drums, sankos, horns and various musical instruments, cased in gold . . . the breasts of the Ocrahs [soul-washers], and various attendants, were adorned with large stars, stools, crescents, and gossamer wings of solid gold.

The stool sat beside the king on its own throne. It was not allowed to come into contact with the ground, nor was anyone allowed to sit on it; even during coronation ceremonies, the king was lowered towards it without ever touching it. In 1900, the British governor, Frederick Hodgson, said he wanted to sit on it – so the Asante hid it. Both sides took up arms, and that was the start of the six-month War of the Golden Stool. Back in London, Lloyd George complained: 'Hodgson's quest of the Golden Stool was something like the quest of the Holy Grail.' The Golden Stool was finally seized by the British and the Gold Coast secured as a British Crown Colony.

The indigenous peoples of north-west America had always known about the gold in Yukon, but it was of little use to them, as they traded in copper. When American prospectors 'discovered' the gold in 1896 they set off a stampede that drew in one hundred thousand people. Fifty years earlier, three hundred thousand had flocked west on news of Californian gold. Something in the American settlers made them crazy for gold.

Gold had taken over from silver as the arbiter of global exchange. From 1699 until 1727, Isaac Newton had been Master of London's Royal Mint. Having applied his reasoning to alchemy and the universal laws of motion, Newton produced an illusion of control over another of the great mysteries – money. In 1717, he created the gold standard, by mistake. By setting the value of gold too high, he devalued silver and established gold as the sole measure of Britain's money. He expected its value to fluctuate but it didn't. It remained steady for the next few years, and then the next two hundred. By the beginning of the twentieth century, other nations had pegged their money systems to gold and it led to a period of free trade and unprecedented prosperity. Only the sudden credit required by the First World War exposed the gold standard's cumbersome nature. Governments discovered that they did not need to buy gold every time they wanted to produce money. They could simply produce money.

In 1906, at the invitation of Mark Twain, Maxim Gorky travelled to New York to raise funds for the Bolshevik cause. He saw the faces of joyless people. The city itself was grim. Sooty steamers slid through litter-filled waters. Trains screeched overhead on elevated track. 'Everything is groaning, howling, grating, in reluctant obedience to some mysterious force inimical to man,' he wrote in his dystopian essay *The City of the Yellow Devil*:

It is as if, somewhere in the heart of the city, a huge lump of Gold were spinning at a terrific pace with voluptuous squeals, powdering the streets with the finest particles, which people catch and seek and clutch at eagerly all day long. When evening falls, the lump of Gold begins to spin in the opposite direction, raising a cold blazing whirlwind, drawing people into it so that they will give back the gold dust they caught during the day. They always give back more than they got and next morning the lump of Gold has grown larger, revolving at a swifter pace, and the exultant screech of iron, its slave, is the clang of all the forces it has enslaved, and sounds louder still.

This is Goethe's *Veloziferische* – a vision of the inflationary century to come.

Gorky returned to Russia more committed than ever to work towards a society where money was not worshipped. When the Revolution came, a ban was placed on the gold trade (dentists were granted an exemption). In 1921, Lenin identified gold as the reason for the war that had recently killed 'ten million men and maimed another thirty million'; he predicted that gold would cause another war before long – perhaps between Japan and the United States. But there would come a time, he assured everyone, when the era of private wealth would be over and money be a thing of the past.

'When we are victorious on a world scale,' he predicted, 'I think we shall use gold for the purpose of building public lavatories.'

By the early 1930s, rapid industrialisation in the Soviet Union had instead resulted in a huge deficit in foreign currency. With the gold reserves from Tsarist Russia long gone, the Soviet authorities opened a shop called *Torgsin* where people could come and sell their old jewellery and Tsarist gold coins. Soon there were 1,500 *Torgsin* stores across the Soviet Union. In ports, the shops acted as semi-sanctioned brothels for foreign sailors. In this way the state raised the equivalent of 220 tonnes of pure gold.

Under Stalin, gold-mining increased rapidly and in secret, until the Soviet Union was second only to South Africa among the world's producers of gold, and not one gram was used for public lavatories.

Mid-September and Tbilisi was still hot. Walking along Rustaveli Avenue, I sought out the shade. Down the side of the Georgian State Museum, I found the staff entrance and two guards reclined in swivel-chairs. One was dozing. I showed the other my papers and asked for Dr Irina Gambashidze.

Irina had headed the archaeological excavations at Sakdrisi. Dating from about 3000 BCE, Sakdrisi was the oldest known gold mine in the world. Irina had worked there for many years and with her had worked the team from Bochum – including Peter and others. They'd all fine-toothcombed the site – they excavated, they dug, they brushed and examined. Stone by stone they cleared back-filled tailings, and they experimented – fire-setting, role-playing Bronze Age miners with Bronze Age tools, to try and understand how, how long, how come? The attached settlement was scrutin-ised in the same way. It had a processing workshop, a graveyard and dwellings, and between them – the mine, death and daily life – they offered an unprecedented cross-section of Bronze Age society, when hierarchies and social complexities familiar to us today first emerged. But there was still work to do.

In 2012, a Russian-owned company applied for a commercial licence to start mining at Sakdrisi. There was undug gold in the ground, and they wanted it. They were refused permission because mining would destroy what was left of the ancient site and Sakdrisi was now protected by law as a 'cultural monument'.

But gold is like gravity; there's no escaping its attraction. The mining company collected supporters in the Georgian government. The protection began to be challenged. Some in parliament resigned over the issue. When the public became aware of the threat, there were angry gatherings in Tbilisi and the police were called. A protest camp was erected at Sakdrisi itself. In the end, none of it did any good. The company was given its licence. Big machinery moved in and, with its remaining archaeology yet to be examined, Sakdrisi was destroyed.

Irina came and signed me in. Together we walked through the museum's back corridors. As we passed glass cabinets, lumpy stumps of stone capitals, mammoth skulls and curling posters, I filled Irina in with news of the Bochum team and their work at the copper complex in Austria. In contrast to the globalising trends of bronzisation, Bronze Age archaeology was a small world.

Irina's unit had a large open-plan office and we found a table at one end of it. She brought out a bag of Victoria plums and arranged them on a plate. 'Fresh from the stall. Please . . .'

She was a tall, elegant woman with a soft and slightly distant manner. Her life, I estimated, was divided roughly into two equal parts – the first half Soviet, the second post-Soviet. The tale of Sakdrisi combined them both. Independence and the end of Soviet rule brought new opportunities, and for Irina, collaboration with the Bochum institute was one of these. For others there came the chance to chase money: it was only a matter of time before gold reserves were targeted.

'Sakdrisi . . . Sakdrisi.' Irina turned over the syllables like the words of a song. 'It was fascinating from the start. We only found a tiny amount of gold. A little on the processing floor, dust really, but none in the graves. Strange to find so little. We think the operation was controlled from elsewhere, and the gold was all taken away. It was connected with Mesopotamia and the Middle East. Powerful places.'

I asked her about the campaign to save the site.

She paused; her voice grew quieter. 'Such a stressful time. They were calling us charlatans on television. The Germans were charlatans, I was a charlatan. There was no ancient gold mine – it was all just medieval. The things they said!'

Then she was approached by a government minister. They needed to revoke the protected legal status of Sakdrisi and would she sign a document to that effect?

'Well, I wasn't going to do that. They said to me, perhaps you'd like some help with another project. Would $2m help? I told them, I will not sign. They said – what about $4m? Your department could do a lot of good work with that.'

It was causing her discomfort to recall it. She reached for a plum, put it carefully on a plate and began to cut away the flesh. We'd been joined by another younger archaeologist. Ketevan had also worked at Sakdrisi and she said: 'What they did! How can they be forgiven? Just for their own gain. I'm always asking myself – why gold? Why do people love it? It has no uses.'

'It's a good question,' I agreed. 'Probably the best question.'

'I think if we could find the answer, we would—' her face was suddenly animated, 'we might – understand everything!'

Irina finished her plum and dabbed her mouth. 'So, I had a telephone call from the Prime Minister's office. I went to see him. He was sitting at a big table and all around him were standing these people from the mining company. He pushed a piece of paper towards me and said: "All we want is your signature." I looked around, and all these men were looking down at me and, well, I started to cry a little.'

But she didn't sign.

*

In the far north-west, as far from Sakdrisi as it is possible to get and still be in Georgia (at least what's left of it after Russian annexations), lies the remote and beautiful region of Svaneti. Its high mountain valleys have always been known for their gold. Villagers there still go looking for it, or so it was said.

A couple of days later, with the sun not yet up, I crossed Tbilisi to find transport to Svaneti. Soon I was in the back of a *marshrutka*, one of the private minibuses that shuttle around the country, squashed up against a large man who occupied part of my seat in much the same way, I thought, that Russia occupies Georgia. The bus filled, then the aisle filled with people on stools, and we were ready to leave, bumping away through Tbilisi's swelling suburbs.

It was my first visit to Svaneti. During a number of trips to Georgia in the 1990s, on various assignments, I had reached most of its restive fringes – Adjara, Akhalkalaki, across the separation lines into Russian-controlled Abkhazia and South Ossetia, over the ridge of the Caucasus to Khevsureti, but never Svaneti. It had a reputation for fierce and clannish independence. I was looking forward to seeing it, if only because it would mark the end of this sardine-tin journey. Ten hours.

The west of modern Georgia lies at the east of the old classical world, where it was known as the land of Colchis. Its significance to ancient Greece was largely metals – silver, copper, iron and gold. The skills of metallurgy also came from here, and Prometheus was chained to its high mountains for stealing them. Strabo – geographer, traveller, philosopher – pointed to the mineral riches of Colchis to conjure a picture of worldly bounty, thereby explaining the extreme efforts of Jason to find the Golden Fleece.

One theory of the Golden Fleece goes like this: in Colchis, sheepskins were used to filter grains of gold from the rivers. Hung out to dry, the wool glittered. Until recently, such a method was practised up in Svaneti. The story of Jason itself survives as a classic hero's tale, a narrative template used for plots from *Star Wars* to *Iron Man*, *Lord of the Rings* to *Harry Potter*. Jason's journey to Colchis involves the full set of challenges: perilous rocks, nasty harpies, six-armed giants, ship-crushing rocks, dragon's-teeth warriors and the final trial accomplished through the perennial power of love. At the centre of it all, the ageless motive, the ageless reward, the very essence of worldly endeavour – gold.

Long before Hollywood, in the early Christian centuries, the story was understood to be about alchemy. Jason is the novice who must undertake a series of intricate experiments before he is introduced to the King of Colchis's daughter, Medea. She is the adept, privy to Hermetic secrets, and by means of certain substances and concoctions, she enables him to achieve his golden goal. Another interpretation suggests that the Golden Fleece is an alchemical formula written on sheepskin.

In fifteenth-century Europe, a band of knights called themselves the Order of the Golden Fleece. Their initial cause was to restore the glory of Byzantium, recently fallen to the Turks. Candidates were sworn in by placing a hand on the back of a living pheasant, representing the Golden Fleece. The Habsburgs were its chiefs – Philip II of Spain, and then his nephew Rudolf II in Prague. It still operates today, a thread of the European questing tradition that is now divided between Habsburg and Bourbon branches. It is described as 'the most prestigious and historic order of chivalry in the world' and its members wear a chain of solid gold with a Golden Fleece pendant.

All morning, we skimmed through the plains of Georgia. I watched from my half-seat the flashing-past of parched yellow prairie. In the mid-afternoon, a distant line appeared above it. I leaned forward. It grew larger, undulating into far-off ridges and peaks. We crossed and re-crossed the Enguri river, its waters lazing through wide banks of pebbles, islands of silt flushed from the mountains, the detritus of a thousand spring thaws. Then we entered the gorge and the sky narrowed to a strip overhead. Mist hid the heights.

Svaneti was always isolated, cut off by this canyon. For fifty miles or more, the road wound up towards its high villages, a slim notch of a route that has made defence possible, a physical deterrent to military hordes of Mongols and Persians and Russians alike. So remote was Svaneti that for long periods the value of its gold was irrelevant. The story goes that in the eighteenth century, one of the princely Dadiani family from the Georgian lowlands was visiting a Svan house when he noticed a large nugget of gold by the fireplace.

'How much do you want for that?' he asked.

'It's not for sale.'

'I'll give you a cow for it.'

'A cow? In that case, you can have it.'

Late in the afternoon, the *marshrutka* stopped at a roadside café. I stumbled out and stood on the road to stretch my limbs like a dog. The noise of the river rose from below. The shadowy forest rose above. For a moment above the valley a portion of cloud opened, and in that part of the sky, I glimpsed a piece of rock, a damp-polished ridge. The sudden clarity gave it the sense of somewhere secret, a little piece of another world, and it made my spirits soar.

The next morning, I did not follow up my contacts. I left all the phone numbers I had gathered unrung. As soon as I started on that route, my time would be gone. I wanted a moment for the place itself.

Snow had fallen overnight. In the village they said it was the first snow of autumn. From about three thousand metres upwards, the slopes were pale, an inverse tide-line bright in the morning sun. The two peaks of Mount Ushba rose at the head of the valley, somehow out of scale, like two planetary molars. A single cloud stood over each of them.

I left the guest house and felt the chill of the air sharp in my throat. Beneath the bridge over the Dolra the water was bubbling over boulders, milky with glacial silt. The village of Twebish lay a little way on. It was no more than a loose scattering of farmsteads and precipitous pasture. On the road, I watched an old woman approach and when she reached me, she stopped and leaned on her stick. Chewing on toothless gums, she looked up at me with something between curiosity and rage.

'Good morning!' I said heartily, in Russian.

'Where are you going?'

'Up there.' I waved my hand skywards. 'To Mount Mezir.'

She continued to glare at me.

'Do you know it?' I was hoping for some reaction. 'Mount Mezir?'

She looked at my boots and at my head. 'Where is your hat?'

'In my bag.'

'Put it on!' And she turned and walked away.

At no more than about two thousand metres, Mount Mezir is not, by Caucasian standards, a high peak but it occupies a powerful place in local lore. Mezir is a domestic deity, traditionally represented as a small metal figurine, often in Svan gold. The mountain has layers of sanctity; an old pre-Christian pilgrimage has been recently re-established at its summit.

The route left Twebish on a steep path, past the leaning barrels of ash pollards and up into old-growth forest. For several hours I climbed through the trees, zig-zagging to ease the slope. Deciduous yellow leaves gave way to hardier spruce and soon they were high overhead, filling the space beneath with their fruity sappy smell. A woodpecker drummed. The wind blew soft through the tops. Then the spruce thinned and the ground opened out and at the forest's upper edge was a fringe of maple and birch in seasonal orange. In the grass, like subterranean beasts pressing their snouts up from the soil, were autumn crocuses and the deep blue of late gentians.

Out of the trees and into a celestial light. The glare was harsh and heady. Two horses, fiercely chestnut in hue, eyed me from a distance. I passed a small lake, like a sky mirror. The lake was also called Mezir and was said to have been a gift from God. It is considered sacrilege to swim in that lake, or throw stones in it, or even to make a loud noise around it. I walked around it, light-footed with reverence.

The peak was a little further on, and there I raised my water bottle and drank. Turning slowly, I took in the site. Through 360°, the snowy peaks and ridges revolved like a high and heavenly rim, serrated, rising and falling, enclosing everything inside. The Russian border lay to the north, the Abkhazia border to the west, and to the south was the rocky profile of the Svaneti range. Here Mount Mezir was less a summit than a high point from which to view higher points. I remained there for some time – and the longer I looked, the more I became aware of being at the centre of something, on the stage of a mountainous theatre-in-the-round.

I thought of other places with that centrality. Garrow on Bodmin Moor and its view of the Rough Tor peak, certain Cretan temples all set in similar positions, and a natural complex on the Turkish–Syrian border I visited years ago, where the site is ringed by hills and each of the hills has a temple on it with the name of a different planet. Such sites are imbued with a hard-to-define rightness. Peaks and clifftop views give a sense of dominance, but these places are about observation, about plain wonder and awe.

A small building stood there, squat and stone-built with a corrugated-iron roof tied down with rock-suspended wires. The wind made little metallic creakings and knockings. Dedicated to the Archangel Mikhael, it is described as a chapel. But it's not really a chapel. It has nothing to do with the Gospels or scripture or the abstractions of organised belief, and everything to do with its position, the solid mountains that surround it. It's about the earth as it is, not what we make it.

The next day I scrolled through the names and numbers I'd collected and chose Narkis Nigoriani, a teacher by profession who had worked in local government. She lived not far away in Lenjeri. Medieval defensive towers rose from the village; large wood-clad dwellings were arranged beneath them. Narkis's home was a collection of buildings around a courtyard. She shared the compound with her extended family – husband, son, son's wife, grandchildren, relations on perpetual visits. She also had rooms for rent.

That evening we sat in her *machubi*, the room for visitors, with slate walls and the woodwork of her husband Salwa – chairs and tables and intricately carved panels and doors. Narkis proved to be a proud Svan who had spent her life promoting and protecting her land. I asked about the language.

'It's very ancient. Svan is one of the world's oldest languages. It's related to Georgian – people say they split apart maybe three thousand years ago. Svan has four hundred Sumerian words in it. The name 'Colchis' is from a Svan word. It means excrement. I don't know why. Maybe the smell of the river.'

There'd been a recent attempt by the Chinese to open a gold mine in Svaneti.

'The Chinese?' Narkis's voice became suddenly louder. 'Yes, yes. They came. We booted them out! We didn't want them here.'

'Why not?'

'We didn't want them spoiling our nature. In Kvemo Kartli, down in Georgia, they have mines and their streams are full of heavy metals. They contaminate the milk and cheese. We don't want that. We want our water clean!'

Ethnic protection and environmental concerns overlapped on the issue, enabled by Svaneti's physical isolation. Narkis had played a part in the booting out, lobbying both those in power and the public. I made a mental note to stay on her good side.

We spoke about her family and she told the story of how her parents met. Her mother had been Jewish, her father Svan. As a teacher, he had been sent to work in a school in Nalchik in the north Caucasus, and it was there he first spotted her mother. He was immediately attracted to her and undertook a courtship, Caucasus-style. He had her kidnapped.

'She could have run away but my father was a very interesting man. So she stayed with him.' Narkis was born a year later. They moved to Svaneti when she was two. 'They said I cried and cried when I left. I didn't want to leave home. Home!' Behind her glasses, her eyes glittered with irony. 'I didn't realise I *was* coming home.' If Narkis was fifty per cent Jewish, she was also a hundred per cent Svan.

A friend dropped by and pulled up a stool and sat in the middle of the room. Vakhtang Pilpani was a large and engaging man. He wore a grey felt cap and talked with the same gusto as Narkis. The Pilpani clan was big in Lenjeri and around Svaneti.

'When I was young,' he recalled, 'we'd have thirty-two of us sitting at the table to eat.'

The Pilpanis had long been musicians and Vakhtang had his own troupe of traditional polyphonic singers. He had toured Europe. He felt lost in Paris, but the small towns were better. 'The people were more friendly – and in some places I was able to see mountains.'

He and Narkis lapsed into Svan. It was conversational at first, but their voices started to rise. The exchanges became shorter, bouncing off each other with ever greater speed, fuelled by quick curses, the whole thing mounting to a crescendo. Just when I thought they might hit each other, they both dropped their shoulders.

'I'm sorry.' Narkis let out a sigh: 'Politics.'

'It's hard to agree sometimes,' I said.

'We *were* agreeing!' cried Vakhtang.

'Up to here.' Narkis tapped the bottom of her chin. 'Our politicians are corrupt right up to here.'

I swung the discussion back to gold, and Vakhtang explained: 'If there was gold in a cliff, or in a mine, they would find a boulder, and set it on a chain and swing it against the rock to break the ore away. Also in the rivers, they used . . . it was like a long box. Here . . .' He took a pencil and made a diagram. As I watched it take shape, I saw it was a sluice – the same design I'd used in Cornwall to separate tin.

'And fleeces?'

'Yes. They used to put sheepskins in the water.'

Salwa, Narkis's husband, came in, carrying a baby granddaughter on his hip. After greetings, he sat cooing and dandling her on his knee. He too wore the Svan grey felt hat; some men even slept in them, though they're taboo for women.

Vakhtang continued. 'You cannot use a machine to get gold. It is forbidden in Svaneti. A few years ago, a man found gold high in the mountains. He took a digger to pieces and flew the bits in by helicopter.'

'The government stopped it,' interrupted Narkis.

264

'So they do some good things?' I said.

Narkis scoffed.

Vakhtang leaned towards me. 'There was a time when the Chinese came and tried to mine gold but—'

'We booted them out!' said Narkis. 'The whole of Svaneti rose against them. Blocked the roads. People lay down in front of their vehicles. They left. They never even got one gram.'

One of the centres of gold in Svaneti was the village of Ieli. I had the name of a man there, and Vakhtang's daughter came with me to help translate and overcome any mistrust of outsiders. Eka Pilpani was in her final year before university, smart and urbane and an enthusiastic reader of crime fiction in English.

When we got to Ieli, we asked around and a young girl pointed up a rocky lane. 'He lives up there.' We knocked on the gate, and a small door in it opened a fraction. The scarved face of an elderly woman appeared. 'Who are you?'

Eka told her.

'You're Pilpani?'

The door opened a little further. One by one, the woman trotted out names and Eka embellished them: 'Baby boy . . . about to be married . . . moved to Tbilisi.' The door opened fully and then the woman explained we could not see her husband because he was very ill.

She walked with us to the edge of the village. She pointed down into the gorge where, beside the pale water of the Enguri river, a defensive tower rose from a cluster of buildings. 'That's where they tried – where they came for gold.'

A man took us down to the river to show us the site. The Enguri here was fast and shallow. Broken trees bristled from a narrow strip on the far side while here the slope was steep, as if in a hurry to escape the waters.

He turned to point back up the hill, at the road we'd just come down. It twisted back and forth, past Ieli and up towards the ridge about a thousand metres above. 'That day the cars covered the whole road. A snake – up there the tail, down here the head. So many cars! It was the first day – and the mining hadn't even started. The Chinese company left that morning.'

He took us to a high bank of alluvial gravel. The mining company had brought bulldozers and flattened part of it for a work-space. At one end was a piece of abandoned machinery – a gravity separator, its hopper orange with rust. 'There were four when they started – the river washed the others away.'

The river's power and Svan pique combined to see off the project. In the running battle between extraction and nature, nature had had a victory. But in the wider war, mining has the edge. Gold's value has risen since the dollar was decoupled from it in 1971, from $35 per ounce to more than $2,000. Gold-mining is now deemed 'one of the most destructive industries in the world'. Tailings dam failures, mercury and cyanide poisoning and tunnel collapses kill thousands of people a year while the production of gold results in an annual 180 million tonnes of waste. The Svans did themselves a favour with their collective action.

We walked back up and stood outside his gate to say goodbye. A black pig was snouting in the dirt. Beyond it slumped an old Kamaz truck, with two wheels missing and a smashed windscreen. The man lived in the only house, behind a high wall, with the old tower proud above it. 'We put a new roof on it two years before.' It was a lone piece of good order.

In the Soviet period, he said, it was all different. That was the golden age. They used to have a bustling market here. People would come from a long way away – from Mestia, from lower Svaneti, for wood and fruit, milk products and Svan salt.

'And gold?'

'Gold too.'

'Do you still look for it?'

'Not so much now. There is a place people go – up there.' He pointed across the gorge, to a high valley. 'No-one gets rich.'

As we left, he turned and pointed downriver: 'In Tskhumari – down there, they still get gold. There's a man called Mackenna.'

'That's not a Svan name,' I said.

'Not Svan,' he scratched his beard and half-smiled. 'I think it's after that film. The Mackenna film.'

Mackenna's Gold. Of course! I'd loved that film. It came out when I was about eight or nine, at the height of my rock-collecting, and its story had something of the same appeal. I must have seen it two or three times. It starred Gregory Peck. The narrative hinges on a secret desert gorge, guarded by Apache spirits, in which there is a vast seam of gold. A group of adventurers are trying to reach it. The plot is immensely complicated and went way over my head. What I remember was the simple sense of trying to reach somewhere difficult, the pull of the gold – and two scenes.

In one, they arrive in the gorge. The far wall glows with an improbable shimmer. There is then a lot of score-settling and killing and a kind of earthquake and the seam of gold is buried. The second scene is one of the last. Gregory Peck – Mackenna – is just about the only survivor. As he rides away, his saddlebag is full of gold nuggets. That redeemed everything.

The film was a classic western – but more than that, it was a classic quest, with structural overtones of the Golden Fleece. It had a celestial cast, a hefty budget and in Carl Foreman a writer-producer whose last western was *High Noon*. But it was a failure. The *New York Times* called it a 'stunning absurdity'. Gregory Peck himself said: 'It was a terrible western. Just wretched.' It was a flop too at the box office. Somehow though, in the stasis of Brezhnev's Soviet Union, the film struck a chord. Sixty-three million people went to see it – the fourth most popular film in Soviet history. It became even more popular in India, watched by more people than any Hollywood film until the coming of *Star Wars* and *Titanic*. My nine-year-old, stone-questing self was with all of them.

Back with Narkis that evening, we made some phone calls. Someone said Mackenna was going to be in Mestia – Svaneti's capital – the next morning. I went to try and find him.

It wasn't hard. A few questions to the group of drivers touting for passengers in the middle of town, a pointed finger, and – waiting patiently inside the Tskhumari *marshrutka* – Mackenna, a sack of flour at his feet.

I stretched out my hand. 'I'm—'

'I know who you are.' He made space on the seat beside him. My asking around had got back to him. 'We shouldn't have to wait very long.'

Mackenna – grey felt cap, kind and elderly face, boyish manner and a tenderness that ran soul-deep. We exchanged the basics using my dodgy Russian, and then the driver winked to me and pointed to the bottles of *chacha* in the kiosk. He gave his throat a flick: 'Come on! Hundred grams?'

I said, no thanks – the roads were bad enough with a sober driver.

I asked Mackenna about the film.

'When I was a boy, it had just come out. I used to go to the river with the old men when they looked for gold. I spent every time I could there. I even dreamed gold. That's why they called me Mackenna. I hardly even remember my other name.'

It was late afternoon before we drove away from Mestia, the minibus full of returning villagers. We left the main road and followed an unmade track down over the Enguri. Then we bumped slowly up to the village, which was a long collection of houses built on a natural step between the gorge below and the forest above. Atop it all was a line of snowy peaks. The track became bumpier, the bus slower, and then Mackenna tapped the driver's shoulder. Together we heaved down his supplies – sack of flour, bags of sugar and tea.

Mackenna lived with his sister, Lola. The two of them were still in the house they'd been born in some seventy years ago. It was a large and spacious house, and there was another smaller one in the garden, half-ruined and creeper-hid. Two more siblings had left Svaneti years earlier and had families of their own in the lowlands. Mackenna and Lola had stayed. They occupied just a few rooms on the ground floor. Beyond were curtained doorways, and from upstairs a ghostly emptiness.

Lola had taught English and German in the village school. She recalled a little English. 'Robert Burns. I liked Robert Burns, popular songs . . . yes, Scotland.' She spoke slowly – and moved slowly, as if time had stretched itself out and she was reclined on it. In her pale face was the shadow of an unnamed illness.

She was a poet too. She rummaged in a bookshelf and took out an anthology of her work. 'They say Mestia is the "throne of poetry". I have published under different names. Look.'

'Will you read one?'

She looked down at the book. 'No, I can't. It's . . . it's not good.'

'I'm sure it is,' I said.

She ran her finger over the cover. 'The name here – "*lemshveni-era*". It means, "It's beautiful."' She opened it and read. 'This road . . .

will not lead me anywhere . . . I will be a thrown stone, heavy – I will be a footprint, washed away.'

I nodded encouragement.

'No, it is not good.'

'It is – *lemshveniera*!'

She turned a few more pages. 'I couldn't open the sky – because it's the sky . . . and my soul is filled with twilight thoughts . . .'

She shook her head. 'Who is that? Is it me? . . . I don't know. It's so sad.'

Mackenna's friend Rezo came round later. Tomorrow they would be going down to the river to sluice for gold. 'The water was high last week,' said Mackenna. 'But it's dropping now. You cannot find gold when the water is high.'

Rezo was a stocky man, without Mackenna's gentleness. He poured three glasses of *chacha*. We drank to families – he had six children, three here with children of their own but the other three had gone, left. Then we drank to women and we drank to the dead and then it started again. The ritual was familiar to me, as was the threshing-floor taste, the flavour of a hundred Caucasian scenes from years ago.

Later, at dusk, I went out to clear my head. Above the village, I gained height fast, following a path up through the forest until I reached a logger's track. The sun had gone down behind the mountains and there was still light in the sky. The cold brought clarity of sorts.

The track dropped and doubled back to face down the valley. In its V was the village, in shadow, a few buildings visible. The medieval defence towers, which in other places had been restored, had collapsed here. The houses they were attached to were generous in scale, large wood-clad buildings with long galleries on the first floor. They were put up largely during the Soviet period – a time that was good for these mountainous areas. Families were big, produce was plentiful, choices few. Now some of the houses were inhabited, many were empty – several, it was hard to tell.

In the 1990s, I'd spent months in such places – in Armenia, in Nagorno-Karabakh, in Georgia and in the mosaic of republics of the north Caucasus. What had I been looking for? Another version of the world, peoples and places off limits for seventy years, stories hidden in the mountains. The Caucasus is one of the world's most culturally diverse areas – languages, conflicts, stories, all thrived and multiplied in its steep-sided valleys. I was amazed by it all. But now? It just looked sad.

Above the village, the high slopes still had the sun. Everything up there was washed in it – the forest, the ridge, the pale glacier above. Flecks of pink-edged cloud filled the evening sky and suddenly I felt cold. I did up my coat and carried on back.

In the morning, we prepared for a day at the river. 'I think I will rest today,' Lola announced. I had the impression she rested most days. I went with Mackenna out into the garden and up the outside steps to the top floor. There was a verandah on one side, facing the mountains. The sun was not yet up behind them, but you could see the bright place where it would appear.

Mackenna looked up and nodded. 'Won't be long now!'

At the far end was a large room with three iron-framed beds around its edge. The mattresses were deeply valleyed. Who had slept here last? In what happy Soviet summer? On one of the beds lay a small chainsaw and a gun. A lightning-streak of a crack dropped down the wall above it, with an icon of Madonna and child hanging lopsided. There was a table with an oilcloth and a collection of old pottery, recovered from the river, and a pair of Zenit binoculars.

Boots, over-trousers, stick, pan . . . Mackenna collected what he needed. He was unhurried, slipping easily into a routine he'd been following for much of his life.

We walked down through the village, past the wooden houses, through a picket fence and across cattle-poached pasture. We paused beside an artificial pond. 'Delicious fish,' said Mackenna.

We traversed a steep bank. At the bottom the ground flattened out again. In places the grass had been cut, but where it stood was blue chicory and yellow trefoils.

'You think Svaneti is beautiful?' asked Mackenna.

'*Konyeshno!* I certainly do.'

We continued towards the gorge in contented silence.

A dog started barking. Golden-coated, high-tailed, she came bounding and bouncing from outside a wooden hut. Rezo was there, scything. He handed Mackenna the scythe and went into the hut. Mackenna dropped the grass in long slow arcs. When Rezo came out of the hut, he had a hatchet and a piece of metal grille, and he was smoking. He watched Mackenna for a moment. 'I have to clear this, get it all off before the snow comes.'

Then he sat on a log and started hammering the grille. 'For the river. It lets the water wash off the mud. Any gold will be underneath.'

We carried on, three of us now, and the dog.

I could hear ahead the surf sound of the river. It grew louder with each step. We dropped down a track that cut through hazel woods to the water's edge. Mackenna looked upstream and downstream.

'Good level?' I asked.

'Not too bad.' He was already looking for the right rocks. I noticed in him fresh energy, as if the river washed the years off him.

Rezo's dog lay down to watch us, chin on her paws. Rezo sat and lit a cigarette.

The milky water before us eddied and crested. Its quick surface was wrinkled with the force of it, and in places there were back waves that rose up briefly and flopped down. A group of midstream boulders cleaved the flow and each had boughs and branches bunched up against it. I stood looking at the river, mesmerised by its endlessly shifting shapes, its deep unseen obstacles, its sheer urgency. It was not at its highest, not even close. Flood detritus lay a long way up the opposite bank.

We began work in the bedrock, an outcrop of dark shale sticking up from a small beach. The bedding planes were vertical and in their folds and nooks, silt had been deposited. With an iron bar we hammered them open, scooping the slatey stuff into Mackenna's pan. They call these places 'bears' kettles', where the heavy gold collects in hollows or cracks. Mackenna started the sifting.

He was standing in the shallows, poised over the pan, a sportsman returning to familiar action as he adjusted the level, spilled the waste and picked out smaller and smaller stones. I looked over his shoulder as he worked, past his stubbly cheek. He was expertly tilting and tipping, the water washing off another layer with each pass. Now the pan's aluminium was visible. The curve of black sand was shrinking in the upper half and then – a single sparkle, a yellow flash. Then another, and another.

I had not expected it so quickly, nor for it to have such impact. These tiny shards of light, flashing grains of cosmic energy! Is the beauty in their value – or is it the other way round? Somewhere in their silent gleam is the answer to the gold enigma – the Pyramid hoards of ancient Egypt, the Klondike craziness, the tasks of the Golden Fleece, all the abstractions of the international market. The pulse quickens. The dead rock gives life. The flakes are revelation, stars in the night sky, very tiny and very large.

A couple more pan-fills, a few more flecks.

'You try.' Mackenna left me to sift a pan. It was a lot harder than it looked. I yielded a few grains only – and felt a familiar excitement, from many years ago.

Mackenna had gone downstream. I could see him stepping carefully through the pebbles, the river rushing around his feet. A house-sized boulder stood before him and he stood before it, as if waiting to gain entry. I saw him reach up to a ledge, and on it was his wooden sluice. He came back with it under his arm. Rezo had started hammering the grille again, bending it to fit with the back of his hatchet.

I went with Mackenna to collect ferns for a filter.

'Do you ever use fleece?'

'We used to.' Mackenna was plucking bright green fronds of bracken. 'But not so many sheep now. Too many wolves.'

It took a while to set up the sluice – and a while to shovel the silt. We took turns, scooping alluvium from the shallows, then tipping it into the chute. The flow was swift through it and it swept away the finer sand first, leaving small pebbles.

Then we took the sluice and removed the grille, and washed out the fern into the pan. More tilting. A few more flakes of gold. With the pad of his finger, Mackenna picked them up and dropped them into a pill jar. 'A little, you see.'

We did a couple more sluices. But the yield was less. I couldn't help seeing a pattern – you find some gold, you finesse the extraction, scale it up. You work harder, shovel-load after shovel-load, and end up with more, certainly. But something is gone, some spirit is lost: the first finding of it. You're already thinking how to improve the searching, how to get more.

We sat down for a rest. Rezo lit a cigarette. Mackenna had put the day's findings in a small clear medicine bottle and filled it with water. The gold made a glinting sediment. The river was a constant sound all around us. We stayed there a few minutes without talking, watching the fast water.

'There used to be more than this,' Mackenna said. 'In the Soviet times they used machines and now it is not so much. My father said there were times they found a stone of gold this big.' He made a golf-ball-sized O with thumb and forefinger.

'But you carry on,' I said.

'Oh, I couldn't stop now.'

We continued with our sitting and looking. The river absorbed us – its deep hiss, its ceaseless rushing. Nothing was still here; everything was on the move. Where the slopes rose from the water, they were steep and easily undercut. In times of flood, in the spring snow-thaw, they would tumble, lumps of soil dropping to the water, to be sifted and rolled and broken again. Birch trees had spidered their roots into the rock, but they too would fall and the boulders would lose their bedding and roll into the water – and all of it repeated, season by season, section by section until far downstream the river came out of its gorge and entered the plain.

Gold was everywhere here – not just in the sands and rock-pockets and the bears' kettles. It was in the soil and the scree and the outcrops, tiny amounts. If you could see it, the whole place would glitter, and if every speck of gold shone like that, it would look like a galaxy from outer space. In the earth's core is a vast hoard of gold – enough, it is said, to coat the entire surface of the planet with a thickness of forty-five centimetres. That would make it worth nothing; it would be litter, pollution. 'The bottom of the world is gold,' wrote Jack Kerouac, 'and the world is upside down.'

I took the medicine bottle and held it up and shook it a little. It flashed in the sunlight.

'It's not pure yet,' said Rezo.

It was pure to me. In this valley, with these men, it was precious and glinting and river-fresh, beauty rinsed of all the madness.

In the early evening, we climbed back to the village. We came up out of the gorge and into the meadows. The last of the light stretched out across the grass, gilding the ground ahead of us. The shadows were long, but we walked in the warmth of the sun. Mackenna ducked beneath a fruit tree and picked me an apple. I bit into it and the juice filled my mouth and he took one for himself and said: 'These ones – they are the best apples in Svaneti.'

Just in front of us, Rezo's dog suddenly went leaping ahead. We watched her go, all fizz and intent, and we smiled as she lolloped through the grass and the autumn flowers, possessed by a nameless energy, by the joy of some imaginary quarry.

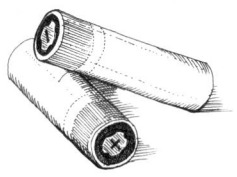

11

LITHIUM

A Coda

IN THE BEGINNING, AFTER THE big bang, there were just
three – hydrogen, helium and lithium. Only when these super-
light primal elements coalesced into stars were the other elements
formed, in nuclear reactions and supernovae. The least dense of all
metals, lithium has the weight of pinewood. You can slice it like
butter. If you place it in water, it will float and then skid across the
surface in a highly reactive death dance.

Scientists have recently discovered some startling aspects of
lithium. Under pressure, its crystalline structure does not break,
but transforms into a substance where different bits have different
properties. The findings have burst open a whole world of possibilities, counter-intuitive chemistry with something of the wild hopes
that alchemists once invested in the earth's materials. Lithium also
promises to be a magical ingredient in next-generation electronics
and quantum computing and in the quest for the Golden Fleece
of energy generation – nuclear fusion. Lithium-ion batteries have
already revolutionised phones and cars.

Since it was first isolated in 1817, lithium has been springing surprises. For most of that time its applications have been medicinal. By dissolving build-ups of uric acid, lithium was found to palliate gout. Naturally occurring in certain waters, it provided a sense of well-being for those who drank them. Lithia Springs in the state of Georgia had long been a Cherokee site of sacred healing before becoming, late in the nineteenth century, a popular spa resort. In the Sweetwater Park Hotel, guests included Mark Twain, Vanderbilts and five US presidents. Lithium water was widely marketed. Lithia beer became popular, 7Up was originally 7Up Lithiated Lemon Soda, and we do not know what crisis or ennui it was that caused President Cleveland in 1895 to write from the White House: 'Send me a case of Lithia Water as soon as possible.'

Now lithium's main medical application is for bipolar disorder, for which it is regarded as the gold standard of treatments. Although it's more effective than synthetic drugs, lithium is strangely under-prescribed – in part because of the risk of toxicity and in part because, as an element, no-one can make money by patenting it. Studies show that lithium can reduce the probability of suicide for bipolar patients by a factor of up to ten. In those parts of Texas, Japan, Austria and Greece where more lithium is present in the water, suicide rates are markedly lower. Lithium has also been effectively administered to treat the negative aspects of chemotherapy, and new research has revealed it as a neuroprotective, offering hopes of treatments for Alzheimer's and Parkinson's. Despite all this, the ways that lithium acts on the body remain a mystery.

High in the Bolivian Andes is a vast and ancient lake whose waters have evaporated to leave a white crust of minerals. Salar de Uyuni has an area about six times that of London and contains the world's largest deposit of lithium. The indigenous Aymara people, for whom the salt flats are the breast milk of the mountain god Tunupa, have protested against the extraction of lithium. Salar de

Uyuni is part of the 'lithium triangle', which, with similar sites in Chile and Argentina, contains over half the world's lithium reserves. In each of them local communities are in opposition: *No al Litio!* Processing of the brines contaminates soils and rivers; to produce a tonne of lithium requires half a million litres of water, and the triangle is one of the driest spots on the planet. Nor do lithium's revenues bring any local benefit. Lithium is an example of the 'resource curse': having precious resources beneath your feet often brings ruin.

Just a few hours' drive from Salar de Uyuni is the city of Potosí. As regional capital, it has been the scene of violent protests against lithium extraction. Above the town rises the treeless mountain of the same name – the bald pyramid known both as 'Cerro Rico' and 'the eater of men'. Silver is still taken from the mountain, which stands as a natural monument not to wealth but to the terrible costs of extraction. Mercury used in ore processing has poisoned the soil for miles around. Inside the mountain, centuries of mining have made the entire structure unstable; working it has become more and more perilous. The area is now one of the poorest in South America.

It was Potosí silver that helped kick-start the age of speed, the age of *Veloziferische*, when ever-faster locomotion began to grow into a collective obsession – from horse power to steam power to hydrocarbons, and now lithium. Here at its source are gutted galleries and destitution.

Back in Cornwall, after the snowy peaks of Svaneti, I found summer still at large. On a warm day in late September, I went to call on my friend Chris at Cornish Lithium. The way there led across the Fal and up towards the Carnmenellis pluton – that geo-tumescence of mineralising granite that in the Bronze Age, and again in the Industrial Revolution, altered the world with its metals. Now it's scabby land. Abandoned shafts and ruined engine houses dot the skyline; in the area around the old mines, only gorse and heather have managed to colonise its toxic grit.

But something is stirring. For the past couple of years, exploratory drills have again been driving into the rocks around Carnmenellis, searching for the deep capillaries where hot lithium-rich brines course through the rock. The drilling has also produced a nascent industry in geothermal energy. At the same time, on the western edge of Carnmenellis, in the last Cornish mine to close – South Crofty – they have begun dewatering its deep levels. It's just possible that tin – the 'darling metal' – may once again be raised from Cornish soil.

Among my mine enthusiast friends, the muddy-kneed explorers, the Beardy Weirdies, the archaeologists and historians, I'd noticed a new excitement. 'Heard about Crofty? They got a pump down Cook's Kitchen.' It was as if after a long period of sleep the earth was waking. All of what Cornwall has become, with its forced dependence on visitors, is now offset by the faint whiff of the old firm, the old trade – pulling treasure from the ground.

I'd first visited Cornish Lithium a couple of years before. It was all quite basic then. The company had just acquired an acre or so of an industrial estate, itself occupying the site of an old tin and copper mine. They'd brought in a Portakabin for an office and a couple of pre-fab sheds. Chris had explained how they'd reached this stage: 'In 1864, there was mention of lithium from a mine near here. It was vague, but enough to go on.' He'd then set out on two years of research through private archives, county records and satellite images, collating mine maps and reports, instances of hot water at holy wells – everything that might offer clues to where lithium might be. He showed me the result on a screen – a 3D underground cross-section, a digitised merging of all the information, zooming in and out on colour-coded shafts and adits.

In those early months, they'd set up a drill right beneath the industrial estate. Beyond the Portakabin, Chris had shown me a waist-high tripod of scaffolding poles, strapped together around a slender shaft. It looked like something a dairy farmer might rig up in a drought. They'd just reached a depth of one thousand metres. In one of the sheds was a rudimentary processing plant – a latter-day alchemist's set-up of pipes. He pointed out a canister of filtration crystals. 'A company makes these to a secret formula. They're pricey. Otherwise—' he waved his hand at the apparatus, 'we got most of this from B&Q.' Two days earlier, a crowd-funder had raised £6m.

Not a lot had changed in the two years since. The same Portaka-bin, the same sheds. The drilling rig was a slightly less shabby tripod and beside it was a stack of plastic drums of brine, waiting for ana-lysis. More widely, lithium had climbed even further up the rankings of desirable resources, with an awareness that its use in electric cars might define the coming age. And things were about to start moving.

'We've been working at twelve other possible drilling sites. One of them is about to go into commercial production. I can't say which.'

We sat around a table in the Portakabin with a couple of others. On the wall were framed facsimiles of the original 1864 mine maps. As we talked, I picked up a strong sense of 'message', of well-rehearsed points put to investors to win their backing. Yet there was also something in their enthusiasm that was genuine, an awareness of something much broader than commercial enterprise.

Lucy was sustainability manager. 'Everything we have nowadays – everything is either grown, or it's mined. It's up to us to do it in a responsible way.'

Cornwall's lithium can never compete with South America's lithium triangle, neither in volume nor in price. What it can offer is 'clean' lithium. There's no intimidation here, no land grabs or collateral damage. Extraction does not gobble up scarce water, nor does anything need to be transported around the world for processing or assembly. Ethical branding of metals is something new, something that highlights the deadly methods of parts of the mining sector, and it's the card Cornish Lithium are playing.

'We offer a version of the circular economy,' explained Chris. 'After the lithium has been taken from it, the water is incredibly pure. We aim to sell that back to South West Water. And the brines come up hot and we can use the heat. We might have glasshouses around the drill. We can grow tomatoes.'

As I left, I bumped into the company's CEO outside. We spoke for a while of general matters, and then he was suddenly articulating the big view. 'For three hundred years, we've been living in an age of fossil fuels – coal first, then oil. This is the future.' He gestured around him – at the Portakabin, the workshop with its B&Q piping, the crude pumping rig. 'The critical materials are things like rare earths and lithium, copper. I believe we are now entering a new age – an age of metals.'

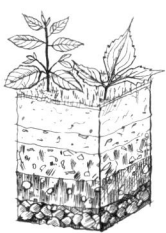

SOIL

THAT WEEK, I WAS WORKING outside. I'd been away for much of the summer and there was catching up to do. I cut the grass in our field and began the task of hand-raking it. It was part of a project to create a few acres of species-rich grassland, and the problem was soil fertility. There was too much of it. The land here had always been a grazing ley, fertilised and improved to ensure the dominance of species like perennial rye-grass and timothy – good for fattening cattle, hopeless for biodiversity. I had planted the hemiparasitic yellow rattle and was committed to reducing excess nutrients by mowing and removing the grass. It was pleasant work in the early autumn sun, but even with a good hay-rake it took forever.

Before the grass restoration, we'd planted seven acres with trees, and left another five for natural regeneration, to see what happened. And what happened was a mass of self-seeded oaks and a lot of sucker-spread blackthorn. I fought against a certain favouritism – the planted trees and the raked meadow were satisfying enough, but I had a preference for the regen, for what emerged without any intervention. As a whole, the old field had taken on a new shape, a scruffy tangle of growth which had been branching out in a dozen different ways, swelling like a capacious idea into something entirely unforeseen. There were not only young trees and meadow flowers but markedly more fauna, more birdlife, more butterflies and invertebrates.

I was astonished at the change and even more so at its speed. We're encouraged to think that the natural world operates at a pace that makes concession neither to the brevity of human life nor to our impatience – forest centuries, evolutionary millennia, geological epochs. Yet each season now brings out something unexpected in the field. Just relinquishing control, removing the old round of grazing and mowing and muck-spreading, was like taking the pressure off a spring – a thousand springs – coiled and waiting in the ground. *Natur Natur sein lassen.*

Working with this small piece of land has taught me many things – about the pace of recovery, about trees and woodland and species succession, about the hegemony of insects, the unsung role of scrub and thorn, the unintended impact of intervention and the value of letting be. But there is one aspect that has surprised me more than anything and that is the soil.

When you begin to look into the soil, it pulls you into a space you didn't even know existed. It's like geological time – or astronomy – but in reverse. To enter its realm you must first adjust your sense of scale, shrink yourself down until earthworms appear as giant beasts, soil megafauna (in good ground, there are about a million per acre, processing two tonnes of organic matter per day). Zoom in further and you find yourself among arthropods, creatures like the two-pronged bristletail, the rotifer and the springtail (the smallest being a fraction of a millimetre, about one hundred thousand of them in a cubic metre of soil). You are in a place now of microtopography, like a mini-mine, of galleries and tunnels and water drops the size of reservoirs, and huge mites scurrying between boulders of sand and clay-cliffs, and wriggling nematodes (accounting for 80% of all animals on earth). Zoom in again to the protozoa (the largest are a hundred microns, one hundred thousand in a *gram* of soil) and there is still one more trophic level, the microbes (bearing a strange similarity to those in our gut). Here, in a single palmful of soil, you are surrounded by ten billion bacteria and eight miles of mycelial fibres. It is a world which, to us big-booted creatures on the surface, is still largely undiscovered. Only about 1% of soil organisms have even been identified.

Beneath the soil in this part of Cornwall is killas, a slatey sedimentary layer described as country rock because it has no useful ores. Yet there is mining here – below the trees, below the grasses and scrub. At every moment of every day, metals are being extracted by plants through their root hairs. No more than a single cell thick, the hairs are invisible to the naked eye but by factoring up the surface area of plants' root systems, they provide a vast platform for the osmosis of mineral salts. In this way iron, magnesium, copper, potassium, cobalt and zinc cross the epidermal threshold into the living realm, into the food chain. The rest of terrestrial life, which is incapable of such mining, depends on this process.

Of the various ways in which human activity has affected the world, few are as shocking as what's happened to soils. Half have been lost in the last 150 years and much of the rest are so depleted, so tractor-crushed, over-fertilised and sterilised by chemicals, that to produce harvests more chemicals and more heavy machinery are applied. Like the fate of Blake's Job, the calamity is one caused largely by ignorance. Hidden beneath our feet, soil was just stuff, a layer of dirt from which crops magically grew. It was there to be manipulated, to be used as a boundless resource, driven over and sprayed and chemically enhanced. It's hard to think it would have happened if its innate vitality and fragility had been understood.

As well as nutrient minerals, plants mine heavy metals. Intensive farming, industrialisation, landfill, mining, fossil fuels – pretty much everything we do at scale – have contributed to a steep rise in soil's toxic constituents. Pulled into plant roots to pass from organism to organism, metals like cadmium and mercury and arsenic do not break down but accumulate, becoming more potent as they climb the trophic ladder towards ruminants, and us. One solution to the problem is phytoextraction – using sacrificial plants to draw out excess heavy metals. Willows have been shown to be good for cadmium and zinc, tomatoes for toxic chromium. Our actions in the world are becoming more and more remedial – planting trees and decontaminant plants, using lithium-ion batteries, intervening in planetary process to offset the consequences of previous interventions. We have entered an age of correction.

So, I was raking the cut grass, pulling it into heaps and removing it. I was trying to correct the former regime's zeal for dosing the soil with stimulants. Rebalancing will allow greater herbal and floral diversity, more pollinators, more voles, more raptors and a generally healthier state in the micro-universe underfoot.

I've found that altering the soil in this way has altered me, correcting my view of the earth. It's refreshed a sense of the physical world as somewhere multidimensional, somewhere whose subtle layers unfold to reveal more layers. To the familiar horizontal picture of my surroundings, I have now added a vertical one. Having seen the land for so long as a map, I now see it also in section – from airborne birds and insects down through earthbound trees and plants and mammals and on into the marvellous subterranean.

The title page of William Blake's central work *The Marriage of Heaven and Hell* shows something similar – a cross-section of the earth. At the top, above ground, are trees and a pale sky and a couple walking together, and that is heaven. Underneath, below the soil, are the flames of hell. But because this is Blake country, nothing is what it first seems. Heaven is not paradise but a sterile place, a place of pure reason and order; in it, the trees stand leafless and bare. Down below, the flames of hell are not punitive flames, but the fires of creative energy and the imagination. Beside them, deep underground, a man and woman embrace, and that is the 'marriage', the fruitful coupling of contrary elements.

Soil ecology was not a thing in the 1790s, and Blake was not overly concerned with agriculture. Yet I found myself reading the image as a schema of the soil. The flames are minerals and metals rising to nourish the roots of the trees, as the imagination nourishes our rationality. In the centre, in the soil itself, Blake presents a series of tiny figures. Look closely and many of them are couples entwined and they're flying up towards the surface, and these too are minerals and nutrients, chemical compounds, and all those micro-organisms are en route to feeding the creatures above. Blake's overarching message in the text of *The Marriage of Heaven and Hell*, and in this image, is the potency of opposites: decay and growth, body and soul, the above and below, the interactive and unifying principles that are at work everywhere and always, in our bodies and in all ecosystems, and in the nano-regions of the soil.

I scooped up the last of the grass, placed it in a waste sack and dumped it among the trees where it would form a patch of habitat. I'd seen grass snakes enjoying the composting warmth. Propping my rake against a young alder, I walked down to the estuary. There's a gap in the hedge where there used to be a five-bar gate, which we still call the 'gate'. Going through it, I came out onto the foreshore, with the feeling it always brings of threshold-crossing, of entering a separate dimension.

The tide was out. The mud stretched away in a flat expanse of neutral tones towards the oak-thick slopes on the far side. The sheer volume of mud is hard to comprehend. It fills the former depth of the valley for miles inland. Once all that mud was soil. Year by year, century by century, the ground upstream was cleared of trees and the soil washed into the river, carried down to where it met the tide, and the load was dropped. Mixed in with it too were the dregs of mining, disturbances of tin-rich alluvium, the tailings of countless leats and sluices, and finer suspensions from china clay dries.

It is now a Site of Special Scientific Interest. A couple of years ago, a marine scientist found a patch of rare intertidal seagrass here, just below the gate. Now there's a large-scale restoration project, sponsored by a clothing brand, to spread the seagrass more widely and protect the vast quantities of carbon stored in the mud. Soil to pollution to precious substrate – all things in time become their opposite.

Later that week, some friends came round for lunch. Afterwards, we wandered down through the trees to the gate. Their two young children went running ahead, skipping down to the foreshore. As soon as they saw the mud, they stopped. For a moment, they stood still and tried to take it all in. I'd seen the same reaction in countless visitors, and remembered my own. As I watched, the two children responded in their chosen way, in a mini-act of human responses to such places, to the raw material of the earth.

The younger one left her shoes on the shingle and stepped onto the mud. She felt it squeeze between her toes. She slid one foot over it like a skater, then the other, and she took a stick and prodded into the cake-like texture, and looked around and found a flat stone to drop and make it splat. She gave a little yelp of joy; she cried out to her brother. He was about eight and he ignored her. He had serious work to do, squatting down to gather the clay into a lump. He found he could mould it into shapes, an orb, a cube, the crude hull of a boat. I watched his absorption, the fascination at seeing forms emerge from something formless, and idly thought of the child all those years ago chipping at the ammonite – and was somewhat taken aback when he turned to ask the age-old question, one that I felt utterly unentitled to answer:

'Please, can I keep it?'

NOTES

1. Ochre

For the ochre at Blombos Cave, see Christopher S. Henshilwood et al., 'A 100,000-year-old ochre-processing workshop at Blombos Cave, South Africa', *Science*, Vol. 334, No. 6053, pp. 219–222, October 2011.

The colour succession in language is explored in Brent Berlin and Paul Kay, *Basic Color Terms: Their Universality and Evolution* (University of California Press, Oakland, 1991).

For ochre's role in human history, see Tammy Hodgskiss, 'Cognitive requirements for ochre use in the Middle Stone Age at Sibudu, South Africa', *Cambridge Archaeological Journal*, Vol. 24, No. 3, pp. 405–428, October 2014, and also her TEDx talk 'Ochre and the dawn of human culture', 15 December 2022, https://www.youtube.com/watch?v=YRPYS4LuKjg.

Recent developments in archaeometric analysis using isotope signatures of tin have meant that archaeological finds can be more easily traced to particular ore sources. The origins of Bronze Age tin have long been something of a mystery, but some firm evidence is now emerging. See Mark Bridge, 'Cornish tin found in Israel is hard evidence of earliest trade links', *Times*, 19 September 2019. Project Ancient Tin is a multidisciplinary endeavour to understand broader questions of tin in antiquity. See https://projectancienttin.wordpress.com. See also R. D. Penhallurick, *Tin in Antiquity: Its Mining and Trade Throughout the Ancient World with Particular Reference to Cornwall* (Institute of Metals, London, 1996).

The quote from Ovid's *Metamorphoses* is from Book I, The Four Ages of Mankind ll. 141–144. Georgius Agricola's *De Re Metallica* (1556; trans. from Latin by Herbert Hoover – later thirty-first president of the US – and republished in 1950 by Dover Publications, New York) gives an overview of the disdain of classical writers for mining, including Euripides, Phocylides, Naumachius, Propertius and Juvenal. As a Renaissance humanist, Agricola strives to overturn the attitude of the times – an 'intensely bitter hatred towards metals', to convince his readers that 'all metals are useful to mankind'. Moral scepticism about the practice of mining was deep-seated in Europe and cited, for instance, by English authors Chaucer, Shakespeare, Milton and Spenser – see George C. Taylor, 'Milton on mining', *Modern Language Notes*, Vol. 45, No. 1, pp. 24–27, January 1930. *De Re Metallica* celebrates and explains the ingenuity of miners in their techniques and was widely used by mining enterprises until the end of the eighteenth century. It draws on Agricola's nine years in the mining town of Joachimsthal (now Jáchymov in the Czech Republic); see Chapter 6, 'Radium'.

The quote from Prometheus is from Aeschylus, 'Prometheus bound', in *Four Plays of Aeschylus*, trans. E. D. A. Morshead (Macmillan, London, 1908), p. 187. For the quote from Hesiod, see *Works and Days*, trans. Dorothea Wender (Penguin, London, 1973), l 174.

The full rhyme about Poldice is: 'At Poldice the men are like mice, / The tin is very plenty / Captain Teague is one of Breage / And he'll give ten for twenty'. See https://www.gwennap-parish. net/gwennap-parish/#:~:text=Historically%2C%20Gwennap%20 was%20the%20richest,effects%20of%20extensive%20copper% 20mining.

For arsenic's history as medicine and poison, see John Frith, 'Arsenic – the "poison of kings" and the "saviour of syphilis"', *Journal of Military and Veterans' Health*, Vol. 21, No. 4, pp. 11–17, December 2013; and Jerome O. Nriagu, 'Arsenic poisoning through the ages', in J. W. T. Frankenberger (ed.), *Environmental Chemistry of Arsenic* (CRC Press, Boca Raton, 2002). For *cantarella* and its use, see Marianna Karamanou et al., 'Toxicology in the Borgias period: The mystery of cantarella poison', *Toxicology Research and Application*, Vol. 2, pp. 1–3, April 2018.

For Paracelsus's maxim about poison, 'All things are poison, and nothing is without poison; the dose alone makes it so a thing is not a poison' ('*Alle Dinge sind Gift, und nichts ist ohne Gift; Allein die Dosis stellt sicher, dass etwas kein Gift ist*'), see Theophrastus Paracelsus, *Dritte Defensio*, 1538. For mention of the Chinese '*du*' and the German '*das Gift*', both of which mean 'poison', see Chapter 8, 'Mercury'. The English word 'dose' comes through French and Latin and the Greek 'dosis' (δόσις), literally, 'a giving'. The earth's 'gifts' represent something similar – benefit or poison, depending on amounts used.

For 'the imaginary world (*univers imaginaire*) . . . discovery of metals' and 'Not only did the manipulation . . . a new mythological and religious universe', see Mircea Eliade, *The Forge and the Crucible: The Origins and Structures of Alchemy*, 2nd ed. (University of Chicago Press, Chicago, 1978), p. 181. The scholarly battle to make alchemy a legitimate subject for study has been most notably fought by Lawrence M. Principe, Drew Professor of the Humanities at Johns Hopkins University. He has published extensively, but the best primer to alchemy and Hermeticism is his book *The Secrets of Alchemy* (University of Chicago Press, Chicago, 2012). I have also had at my elbow, when writing about the subject in this and later chapters, Alexander Roob, *Alchemy & Mysticism* (Taschen, Cologne,

2023); William R. Newman, *Promethean Ambitions: Alchemy and the Quest to Perfect Nature* (University of Chicago Press, Chicago, 2004); and the collection of essays in Ingrid Merkel and Allen G. Debus (eds.), *Hermeticism and the Renaissance: Intellectual History and the Occult in Early Modern Europe* (Folger Books, Washington, 1988).

2. Tin

The theory of the tin buttons of *La Grande Armeé* is so far unproved, and probably unprovable. See Jesse Beckett, 'Did tin buttons cause Napoleon's army to retreat from Russia?', *War History Online*, December 2021, https://www.warhistoryonline.com/war-articles/tin-buttons-napoleons-army.html. However, the deterioration of pure tin below 13.2 °C is well attested. See Ben Cornelius et al., 'The phenomenon of tin pest: A review', *Microelectronics Reliability*, Vol. 79, pp. 175–192, December 2017.

For stanene, see R. Colin Johnson, 'Stanene may be better than graphene', *EE Times*, 12 March 2013, https://www.eetimes.com/stanene-may-be-better-than-graphene/.

The Phoenician link to Cornwall has been debunked by Craig Weatherhill among others. See Craig Weatherhill, 'Where the Phoenicians weren't', *Cornwall Yesteryear*, undated, https://cornwall-yesteryear.com/where-the-phoenicians-werent-by-craig-weatherhill.

For information on the St Mawes ingot in the Royal Cornwall Museum, see https://projectancienttin.wordpress.com/2020/12/22/tin-ingots-artefacts/.

'This darling Metal . . . in the heart' is from William Pryce, *Mineralogia Cornubiensis: A Treatise on Minerals, Mines, and Mining* (1738, reprinted in facsimile 1972, Bradford Barton, Truro), p. ix;

'unremitting ardour ... under the sun', ibid., p. viii. The following is an account of tin fever – not that of a miner but one who searched the Red River for alluvial tin: 'He stuck to it like a leech, up in time in the morning jumping into any passing donkey-cart to get there more quickly, but the fruits of the river soon produced him a pony and trap, and now he has a hoe, trap and silver-mounted harness.' From D. B. Barton, *A History of Tin Mining and Smelting in Cornwall* (Cornwall Books, Exeter, 1969), p. 152.

For James Forbes's account of a Cornish mine, see James Forbes, 'Tour into Cornwall to the Land's End: ... describing the Scilly Isles, the tin mines, Portsmouth, Stonehenge, etc etc. In letters by James Forbes to his wife's sister Ann ... in 1794', *Journal of the Royal Institution of Cornwall*, Vol. 9, No. 2, pp. 146–206, 1983.

An account of the Charter of the Stannaries in Cornwall and the special privileges of the Cornish miners is given in A. K. Hamilton Jenkin, *The Cornish Miner: An Account of His Life Above and Underground from Early Times* (Allen & Unwin, London, 1962), pp. 32, 37.

A full and detailed lexicon of eighteenth-century Cornish mining is given in Pryce, op. cit., pp. 315–331.

For the miners' terms and superstitions surrounding animals, see Hamilton Jenkin, op. cit., p. 52, and for the names of setts, see ibid., p. 68.

3. Peat

'Brighid of the peat-heap' appears in *Genealogy of Brighid*, trad., translated from Gaelic by Noragh Jones, in Noragh Jones, *Power of Raven, Wisdom of Serpent: Celtic Women's Spirituality* (Floris Books, Edinburgh, 1994).

For discussion of the relationship between the distributions of Viking settlements and bog iron, see Graham Bowles et al., 'Viking expansion and the search for bog iron', *PlatForum*, Vol. 12, pp. 25–37, 2011.

'Then the blacksmith . . . in the thickets' is from Elias Lönnrot (ed.), *The Kalevala: The Epic Poem of Finland*, trans. John M. Crawford (The Robert Clarke Company, Cincinnati, 1898), Rune IX. Origin of Iron.

'kind, black butter' is from Seamus Heaney, 'Bogland', in the collection *Door into the Dark* (Faber & Faber, London, 1969). 'Found in a bog' is from Seamus Heaney, *Preoccupations: Selected Prose, 1968–78* (Faber & Faber, London, 1984), p. 54. Peat bogs were multilayered for Heaney as 'a landscape that remembered everything that happened in and to it. In fact, if you go round the National Museum in Dublin, you will realize that a great proportion of the most cherished material heritage of Ireland was "found in a bog"'. The poet, for him, also had something of the peat-digger, the miner, drawing from the ground memories and meaning: 'I have always listened for poems, they come sometimes like bodies come out of a bog, almost complete, seeming to have been laid down a long time ago, surfacing with a touch of mystery', ibid., p. 34. For bog butter, see Muiris O'Sullivan and Liam Downey, 'Bog butter: Why was it buried?', *Archaeology Ireland*, Vol. 33, No. 2, pp. 27–29, Summer 2019.

For the Scottish Gaelic *sùil-chruthaich*, 'eye of creation', see Donald Murray, *The Dark Stuff: Stories from the Peatlands* (Bloomsbury, London, 2019), p. 231. The Irish bog lexicon is from Niall Ó Dónaill, *Foclóir Gaeilge–Béarla*, 1977, https://www.teanglann.ie/en/fgb/.

The case for the Golden Age of the Dutch being attributed to its access to cheap fuel, in the form of peat, is made in some detail in J. W. de Zeeuw, 'Peat and the Dutch Golden Age: The historical meaning of energy-attainability', *AAG Bijdragen*, Vol. 21, pp. 3–31, 1978.

Information on the Cruquius pumphouse and drawings and technical details of the pump can be found in 'International historic mechanical engineering landmarks: Haarlemmermeer Cruquius', 19 June 1991, https://web.archive.org/web/20110725023725/http://files.asme.org/ASMEORG/Communities/History/Landmarks/5525.pdf.

Procopius of Byzantium is quoted in Johan van Veen, *Dredge, Drain, Reclaim: The Art of a Nation* (Nijhoff, The Hague, 1948), p. 8. The quote 'try to warm their frozen bowels … than in the sun, which one hardly ever sees' by Pliny the Elder is from Roger H. Charlier et al., 'Panorama of the history of coastal protection', *Journal of Coastal Research*, Vol. 21, No. 1, pp. 79–111, January 2005.

The story of the Netherlands' long battle with water, from the first settlers through to the medieval period and the engineering projects of the twentieth century, has been widely told. I have drawn on a number of sources, including Frank van Schoubroeck, 'The remarkable history of polder systems in the Netherlands', 2010, https://www.fao.org/fileadmin/templates/giahs/PDF/Dutch-Polder-System_2010.pdf; Petra J. E. M. van Dam, 'Sinking peat bogs: Environmental change in Holland, 1350–1550', *Environmental History*, Vol. 6, No. 1, pp. 32–45, January 2001; William H. TeBrake, 'Taming the waterwolf: Hydraulic engineering and water management in the Netherlands during the Middle Ages', *Technology and Culture*, Vol. 43, No. 3, pp. 475–499, July 2002; and Paul Wagret, *Polderlands* (Methuen, London, 1968).

The poem by Joost van den Vondel appears on a map of the Haarlem Lake drawn by Jacob Bartelz Veris in 1641. It is reproduced on a wall of the Haarlemmermeermuseum De Cruquius.

The article that led me to Jippe Kreuning is his 'The many faces of the Waterwolf, the Low Countries' greatest friend and foe', 2015, http://www.waterinitiativeforthefuture.org/blog, which also offers a good overview of Dutch flooding and the remedial role of windmills.

A detailed analysis of the *St Elizabeth's Day Flood* diptych in the Rijksmuseum is given by Hanneke van Asperen in 'Charity after the Flood: The Rijksmuseum's St Elizabeth and St Elizabeth's Flood altar wings', *The Rijksmuseum Bulletin*, Vol. 67, No. 1, pp. 30–53, March 2019.

Jim Leary, *The Remembered Land: Surviving Sea-Level Rise after the Last Ice Age* (Bloomsbury Academic, London, 2015) is a fine study of the varying rate of sea-level rise post-Pleistocene and its implications for the small groups living in Doggerland. See also Fraser Sturt et al., 'New models of North West European Holocene palaeogeography and inundation', *Journal of Archaeological Science*, Vol. 40, No. 11, pp. 3963–3976, November 2013. The particularities of sea-level rise in the area around the Biesbosch are examined in Marc P. Hijma and Kim M. Cohen, 'Holocene sea-level database for the Rhine-Meuse Delta, The Netherlands: Implications for the pre-8.2 ka sea-level jump', *Quaternary Science Reviews*, Vol. 214, pp. 68–86, June 2019.

Peter the Great's building of St Petersburg is recounted in Alexander Pushkin's epic poem 'The Bronze Horseman: A St Petersburg Tale', which was prompted by the 1824 flood. Translating Pushkin is famously challenging, akin to establishing a city in a bog. Vladimir Nabokov, who rendered the poem in a 'plain, prosy and rhymeless' version, was fiercely critical of another translator who attempted to mirror the Russian style, stating in *New York Review of Books* in 1964: 'The task of twisting some five thousand Russian iambic tetrameters, with a rigid pattern of masculine and feminine rhymes, into an equal number of similarly rhymed English iambic tetrameters is a monstrous undertaking.' Frederick the Great's swamp-draining is explained in David Blackbourn, *The Conquest of Nature: Water, Landscape and Making of Modern Germany* (Jonathan Cape, London, 2006), pp. 41–43. For the story of the Pontine Marshes, see 'Reclaiming the Pontine Marshes', *Wonders of World Engineering*, Vol. 1, No. 2, 1937, https://wondersofworldengineering.com/pontine_marshes.html. For over two thousand years the area has oscillated between malarial devastation and agricultural prosperity.

I was first alerted to Goethe's *Faust* as a parable of progress by Marshall Berman's masterly *All That Is Solid Melts into Air: The Experience of Modernity* (Simon & Schuster, New York, 1982). Goethe combines in *Faust* the 'cultural ideal of self-development and the real social movement towards economic development', ibid., p. 40. For the text of *Faust*, I have used the English translation of Martin Greenberg, *Faust: A Tragedy* (Yale University Press, New Haven, 2014): 'It flows along . . . where, tell me, is the gain?', ll. 10549–10552; 'I'll bar the lordly sea out from the lowly shore!', ll. 10564–10565; 'To drain that pestilential swamp . . . a second paradise', ll. 11919–11926; 'These dikes and dams . . . it's hopeless', ll. 11903–11905; 'To see such life . . . ground that is free!', ll. 11939–11940.

'The soul of Faust . . . strivings' and 'What ended the endless strivings . . . a material mastery over the world' are from Nicholas Berdyaev, 'The pre-death thoughts of Faust' (1922). See 'Notes along the way: The books and articles of Nicholas Berdyaev', undated, https://www.nicholasberdyaev.com/articles/articles-1920-1929.

'Your schemes . . . destruction, ruin' is from Greenberg, op. cit., ll. 11906–11908.

'*Verweile doch, du bist so schön!*' (Linger awhile, you are so beautiful!)', ibid., ll. 11942. I have altered Greenberg's rendition of '*schön*', from 'fair' to 'beautiful', feeling it to be clearer and more direct. Each to their own.

For a historical view of the shifting landscape of the Biesbosch, I used Wim van Wijk, *Historische Atlas van de Biesbosch* (Uitgeverij WBOOKS, Zwolle, 2012).

4. Bronze

The elaborate description of Hephaestus and the forging of Achilles' shield is in Homer, *The Iliad*, Book 18, ll. 478–608. The shield is an early example of a microcosm, an object that aims to contain in it the entirety of things, representing not only the universe but the idea of essential unity. For more on microcosms, see Chapter 7, 'Aerolite'.

The Nebra Sky Disc, like equivalent stone monuments, is an object that invites a colourful range of interpretations. The gold orb on it could be the sun or it could be a full moon; the likelihood of it being the moon is presented in Duncan Garrow and Neil Wilkin, *The World of Stonehenge* (British Museum Press, London, 2022), p. 145. The thirty-two stars and their configuration, including the Pleiades, have also been examined in detail; see Henning Dathe and Harald Krüger, 'Morphometric findings on the Nebra Sky Disc', *Time and Mind*, Vol. 11, No. 1, pp. 89–104, 2018. Doubts were at one stage expressed about the disc's date (in part because there is nothing from the period remotely like it); see Rupert Gebhard and Rüdiger Krause, 'Critical comments on the find complex of the so-called Nebra Sky Disk', *Archäologische Informationen*, Vol. 43, pp. 325–346, 2020; and Becky Ferreira, 'A bitter archaeological feud over an ancient vision of the cosmos', *New York Times*, 19 January 2021. But these have been dispelled; see Ernst Pernicka et al., 'Why the Nebra Sky Disc dates to the early Bronze Age.

An overview of the interdisciplinary results', *Archaeologia Austriaca*, Vol. 104, pp. 89–122, 2020. For the significance of the solar boat, see Andrea Vianello, 'The ship and its symbolism in the European Bronze Age', in Fernando Coimbra and George Dimitriadis (eds.), *Cognitive Archaeology as Symbolic Archaeology* (BAR, Oxford, 2008), pp. 27–34. For the sky disc, I also made use of A. F. Harding, *European Societies in the Bronze Age* (Cambridge University Press, Cambridge, 2000); and Jan-Heinrich Bunnefeld et al., *The World of the Sky Disc – New Horizons* (Landesmuseum für Vorgeschichtem, Halle, 2021).

For the essay on bronzisation, see Helle Vandkilde, 'Bronzization: The Bronze Age as pre-modern globalization', *Prähistorische Zeitschrift*, Vol. 91, No. 1, pp. 103–223, 2016. The quote 'the Bronze Age constituted . . . one crucial resource' is from ibid., p. 103, and 'Bronze Age mobility . . . rare materials' is from ibid., p. 112.

The quote 'an aura of prestige and awe' is from Mary W. Helms, *Ancient Panama: Chiefs in Search of Power* (University of Texas Press, Austin, 1979), p. 176. See also Mary W. Helms, 'Nourishing a structured world with living metal in Bronze Age Europe', *World Art*, Vol. 2, No. 1, pp. 105–118, July 2012, for 'Bronze artifacts . . . beyond the horizon' (p. 108) and 'the sheer human effort . . . the deposited bronze' (p. 109).

For the global consumption of sand and gravel, see Tommy Greene, '50bn tonnes of sand and gravel extracted each year, finds UN study', *Guardian*, 26 October 2022.

For a full account of Gregor Borg's sourcing of the Nebra Sky Disc's gold to Cornwall, see Ines Godazgar, 'The trail leads to Cornwall', *Campus Halensis*: Das Onlinemagazin der Martin-Luther-Universität Halle-Wittenberg, 21 October 2019, https://www.campus-halensis.de/en/artikel/die-spur-nach-cornwall/. See also Anja Ehser et al., 'Provenance of the gold of the Early Bronze

Age Nebra Sky Disk, central Germany: Geochemical character-ization of natural gold from Cornwall', *European Journal of Miner-alogy*, Vol. 23, No. 6, pp. 895–910, 2011. I also drew on private conversations with Dr Borg and Courtenay Smale.

Lines 266–276 of 'The Canon's Yeoman's Tale' in Geoffrey Chau-cer's *The Canterbury Tales* give the full passage of 'the four spirits and the seven bodies'. Chaucer was said to have based his yeoman on a real case of alchemy; see Euan Roger, 'The Alchemist's Tale: Geof-frey Chaucer and alchemy', The National Archives, 9 October 2019, https://blog.nationalarchives.gov.uk/geoffrey-chaucer-and-al-chemy. I also used Yannis Almirantis, 'The paradox of planetary metals', *Journal of Scientific Exploration*, Vol. 19, No. 1, pp. 31–42, 2005.

5. Silver

For the story of Potosí, see Kris Lane, 'Potosí', in Mark Thurner and Juan Pimental (eds.), *New World Objects of Knowledge: A Cabinet of Curiosities* (University of London Press, London, 2019). For the patio process, see Ronald D. Crozier, 'Silver processing in Spanish America: The patio process and beyond', *CIM Bulletin*, Vol. 86, No. 972, pp. 86–91, July–August 1993. For the significance of the Pacific route for Potosí silver and the demand in China for silver, see Dennis O. Flynn and Arturo Giráldez, 'Born with a "silver spoon": The origin of world trade in 1571', *Journal of World History*, Vol. 6, No. 2, pp. 201–221, Fall 1995.

The expansion of silver production in the early modern period is examined in John N. Nef, 'Silver production in central Europe, 1450–1618', *Journal of Political Economy*, Vol. 49, No. 4, pp. 575–591, August 1941. For the economic aspects of the Reformation, see Robert B. Ekelund, Jr. et al., 'An economic analysis of the Prot-estant Reformation', *Journal of Political Economy*, Vol. 110, No. 3, pp. 646–671, June 2002, which explores the 'argument originated by Adam Smith (1776) and formalized by Iannaccone (1991) . . . that state-supported religious monopolies behave inefficiently in many ways, thereby opening up the possibility of entry by more efficient competitors' (p. 647).

The story of the evacuation of the Stolberg-Stolberg family at the end of the Second World War is told by Henry Henshaw in 'The escape: The evacuation of the prince of Stolberg-Stolberg and his family from Stolberg Castle in 1945', *Illustrierte Geschichte, Grafschaft Stolberg*, undated, http://www.grafschaft-stolberg.de/ s010011.

For the general biographical details of Goethe, I have relied mainly on Rüdiger Safrinski, *Goethe: Life as a Work of Art*, trans. David Dollenmeyer (Liveright, New York, 2017).

Goethe recounted his near-death experience down a mine in a letter to Charlotte von Stein on 9 December 1777, writing 'fate paid me a great compliment'. See K. Mandelklow (ed.), *Briefe an Goethe* (Wegner, Hamburg, 1965), Vol. I.

For Christopher Polhem as the 'Archimedes of the North' and feats of engineering in the Harz more generally, see Christoph Bartels, 'Georg Winterschmidt's water pressure engines in the Upper Harz mining district 1747–1763: Plans, experiments, prob-lems, results', *Icon*, Vol. 3, pp. 24–43, 1997. Leibniz had visited the Netherlands before working in the Harz and had watched the use of wind in pumping out the polders, as described in Andre Wakefield, 'Leibniz and the wind machines', *Osiris*, Vol. 25, No. 1, pp. 171–188, 2010. Gottfried Wilhelm Leibniz, *Protogea*, ed. and trans. Claudine Cohen and Andre Wakefield (University of Chicago Press, Chicago, 2008) gives a fascinating view of the supreme seventeenth-century polymath and the formative years he spent in the mining regions of the Upper Harz.

Goethe's optimism on 21 February 1827 – 'The resulting benefits . . . but I won't' was recorded by Johann Peter Eckermann on 21 February 1827. See *Conversations with Goethe: In the Last Years of His Life* (Penguin Classics, London, 2022), pp. 502–503. 'The worst torment / To have so much, yet still to want' is from Johann Wolfgang von Goethe, *Faust: A Tragedy*, trans. Martin Greenberg (Yale University Press, New Haven, 2014), ll. 11607–11608.

For the link between medieval mining and lead pollution, see Joseph McConnell et al., 'Pervasive Arctic lead pollution suggests substantial growth in medieval silver production modulated by plague, climate, and conflict', *Proceedings of the National Academy of Sciences of the United States of America*, Vol. 116, No. 30, pp. 14910–14915, July 2019.

'a transcendence which erases the productive possibilities' is from Susan Stewart, *On Longing: Narratives of the Miniature, the Gigantic, the Souvenir, the Collection* (Duke University Press, Durham, 1992), p. 60.

'Father of love . . . / . . . the thousand fountainheads' is from Johann Wolfgang von Goethe, 'Harzreise im Winter' ('Winter Journey in the Harz', trans. Christopher Middleton), in Matthew Bell (ed.), *The Essential Goethe* (Princeton University Press, Princeton, 2016), p. 10 (translation slightly changed by author).

The quote from Herder – 'At that time . . . Everyone mineralogized' – appears in E. P. Hamm, 'Unpacking Goethe's collections: The public and the private in natural-historical collecting', *The British Journal for the History of Science*, Vol. 34, No. 3, p. 280, September 2001. 'Goethe . . . sublime detachment of the scientific thinkers' is from Charles Sherrington, *Goethe on Nature and on Science* (Cambridge University Press, Cambridge, 1949),

p. 30. For Goethe's early immersion in the literature of alchemy, see Ronald Gray, *Goethe, the Alchemist* (Cambridge University Press, Cambridge, 1952), pp. 3–37. For Goethe's work on the intermaxillary bone, see Ryan Feigenbaum, 'Toward a nonanthropocentric vision of nature: Goethe's discovery of the intermaxillary bone', in Adrian Daub and Elisabeth Krimmer (eds.), *Goethe Yearbook 22* (Boydell & Brewer, Martlesham, 2015), pp. 73–94.

'The world is so great and rich . . . never want occasion for poems', said by Goethe on 18 September 1823, is from Eckermann, op. cit., p. 37. 'I am beginning to grow aware . . . insight and joy to someone' was written in a letter to Frau von Stein on 9 July 1786. 'I worship him . . . That is *my* god!' (20 February 1831) is from Eckermann, op. cit., p. 386. 'I saw one life / and felt that it was joy' is from William Wordsworth, *The Prelude* (1805), Book II: School-Time, l. 430.

'as clear as my face . . . my tears of joy' is from Safrinski, op. cit., p. 206; 'an altar of sweetest thanks . . . / . . . with hosts of spirits' is from 'Winter Journey in the Harz', in Bell, op. cit., p. 11; and 'Here, on this primal and everlasting altar . . . which are so near' is from Goethe's 1784 essay 'On Granite', reproduced in full in Bell, pp. 913–915. Goethe's further thoughts on granite, and its triune structure – *die Dreieinheit* – are quoted and discussed in Gray, op. cit., pp. 133–159. 'Nature! We are surrounded . . . getting more life' is from T. H. Huxley's own translations of Goethe, which appeared as 'Nature: Aphorisms on Goethe', the first article in the first issue of *Nature* in 1869. Goethe's aphorisms were a little poetic for some, including Darwin, who wrote to Hooker as follows: 'Lord, what a rhapsody that was of Goethe, but how well translated; it seemed to me, as I told Huxley, as if written by the maddest English scholar. It is poetry, and can I say anything more severe?', from 'Goethe's reflections on nature', *Nature*, Vol. 129, p. 425, March 1932. In fact, there is now the suggestion that Goethe did not write them in the first place:

> These aphorisms were generally attributed to Johann Wolfgang von Goethe, one of the greatest poets and scientists of the 19th century (in the article Huxley points in particular to his interest in comparative anatomy). However, Huxley notes that Goethe himself wasn't totally sure whether he had actually written the aphorisms or not. Goethe had been sent the text in 1828 because the sender thought it was his sort of writing, and Goethe, looking at it, thought that it was indeed the sort of stuff he could have written in 1786.
>
> It is now thought that the writer of the aphorisms was Georg Christoph Tobler and that they were 'first published in 1783 in the *Tiefurt Journal*. Tobler wrote the essay after repeated conversations with Goethe.'

(So they were, at least, Goethe's expressed thoughts.) See Brigitte Nerlich, 'Nature's first article: Huxley on Goethe', Blogpost, University of Nottingham, 11 January 2019, https://blogs.nottingham.ac.uk/makingsciencepublic/2019/01/11/natures-first-article-huxley-on-goethe/.

'Every plant now declares . . . flower-head can speak' is from Goethe's 1798 essay 'The Metamorphosis of Plants', in Bell, op. cit., p. 26. For a discussion of the *Urpflanze* – 'Goethe's primal plant', see Gray, op. cit., Chapter IV.

'The moment I get . . . rocks and minerals' is from Goethe, *Italian Journey (1786–1788)*, trans. W. H. Auden and Elizabeth Mayer (Penguin Classics, London, 1970), p. 114. Goethe follows this thought with 'I seem to be an Antaeus who always feels new strength whenever he is brought into contact with his mother earth.' Antaeus's mother was Gaia, and Hercules only defeated him when he raised him from the earth and crushed him to death in a bear hug. Goethe's lifelong love for rocks sustained him and gave him the sense that they nourished his spirit. Leaving Rome in 1788 prompted him to write to his friend Karl Ludwig von Knebel, 'tap on the rocks to drive out the bitterness of death' (quoted in Safrinski, op. cit., p. 297).

'Stones are mute teachers . . . cannot be passed on to others' is quoted in E. P. Hamm, 'Unpacking Goethe's collections: The public and the private in natural-historical collecting', *The British Journal for the History of Science*, Vol. 34, No. 3, pp. 297–298, September 2001. 'Sublime tranquillity' is from Goethe, 'Über den Granit', in Bell, op. cit., p. 914.

For Goethe's general writing on geology, see W. Scott Baldridge, 'The geological writings of Goethe: Despite his keen powers of observation, Goethe's ideas on geology reflected the biases of his time', *American Scientist*, Vol. 72, No. 2, pp. 163–167, March–April 1984; Heather Sullivan, 'Collecting the rocks of time: Goethe, the romantics and early geology', *European Romantic Review*, Vol. 10, Nos. 1–4, pp. 341–370, 1999; and Heather Sullivan, 'Organic and inorganic bodies in the age of Goethe: An ecocritical reading of Ludwig Tieck's "Rune Mountain" and the earth sciences', *Interdisciplinary Studies in Literature and Environment*, Vol. 10, No. 2, pp. 21–46, Summer 2003, in which Sullivan discusses the fluidity between the organic and the inorganic among eighteenth-century thinkers like Goethe and the theme of Romantic alienation in the context of the inorganic – the cave, the mine, etc.

'"folding over" of natural phenomena so that things unjoined before now connect in relationship' is from David Seamon, 'Connections that have a quality of necessity: Goethe's way of science as a phenomenology of nature', *Call to Earth*, Vol. 4, No. 1, pp. 3–11, March 2003.

For a discussion of Goethe's neologism 'Veloziferische', see Bryan Norton, 'Veloziferisch (Velociferian)', *Lexicon of Philosophical Concepts*, Vol. 1, No. 1, January 2021. Norton explains: 'Das Veloziferische marks a dangerous speed at which organic growth is outpaced by the rapid acceleration of technological development. At velociferian speeds, the otherwise figurative role of negation in Goethe's philosophy of nature takes on a disfiguring function, highlighted most clearly by the techno-accelerationist allegory *Faust*.' For Marx's view on credit creation and acceleration, see Karl Marx, *Capital: A Critique of Political Economy*, Vol. III (1894), Part IV, Chapter 27.

The identification of Goethe as the 'greatest money theorist' is from Jacques Rueff, *The Age of Inflation* (Henry Regnery Company, Chicago, 1964), pp. 62–63. Rueff goes on to quote from *Faust*, in which the illusion of credit is presented:

> But lo, a trick quite new to me:
> The thing each seizes eagerly
> Rewards him with a scurvy pay,
> The gift dissolves and floats away. . . .
> Some grab, and catch frail butterflies.
> The rascal offers wealth untold,
> But gives the glitter, not the gold.

For an appraisal of the hidden costs of Bitcoin, see Gabriel Dance et al., 'The real world costs of the digital race for Bitcoin', *New York Times*, 10 April 2023. Details of hyperinflation in the Weimar Republic can be found at https://www.johndclare.net/Weimar_hyperinflation.htm.

6. Radium

'fantastic exception, a mad inversion . . . in the constitution of matter' is from H. G. Wells, *The World Set Free* (Macmillan, London, 1914), Preface, Section 8. For 'as if I should now utter piercing shrieks and act like a maniac on this platform', see William James, *Pragmatism: A New Name for Some Old Thinking* (Longmans, Green and Co., London, 1907), Lecture II – What Pragmatism Means.

For the general biography of Marie Curie, I used Robert Reid, *Marie Curie* (Collins, Glasgow, 1974).

The description 'a cross between a stable and a potato-cellar' was by the German chemist Wilhelm Ostwald, as quoted in Albert Goldbarth, 'Fuller', *The Kenyon Review*, Vol. 6, No. 2, p. 106, Spring 1984.

'Unfathomably rare and intensely powerful . . . the modern world had ever seen' is from Luis A. Campos, *Radium and the Secret of Life* (University of Chicago Press, Chicago, 2015), p. 1. For 'newspapers have become radioactive', see ibid., p. 42.

'No other chemical element . . . its commercial production' is from S. C. Lind, 'Radium', *The Scientific Monthly*, Vol. 20, No. 6, pp. 606–611, June 1925.

'It was as if a human being . . . three hundred years' is from Mrs E. M. Simpson, 'Radium', *The American Journal of Nursing*, Vol. 4, No. 7, p. 528, April 1904.

'When scientists discovered radium . . . a revolutionary beauty secret' is from Lucy dos Santos, *Half Lives: The Unlikely History of Radium* (Pegasus, New York, 2020), excerpted from *Literary Hub*, 8 July 2021, https://lithub.com/a-revolutionary-beauty-secret-on-the-rise-and-fall-of-radium-in-the-beauty-industry/. Crème Activa's claim is from the same source. Doramad Toothpaste was produced in Germany in the 1920s and 1930s as Doramad Radioaktive Zahncreme, promising teeth of 'radioactive brilliance'.

For radium tea in Paris, see 'Radium cure a fad of Paris society; patients play bridge and take tea for two hours daily in a drawing room', *New York Times*, 10 December 1911.

The early popularity and uses of radium in entertainment are outlined in Lawrence Badash, 'Radium, radioactivity, and the popularity of scientific discovery', *Proceedings of the American Philosophical Society*, Vol. 122, No. 3, pp. 145–154, June 1978.

'a fungating mass covered with cauliflower excrescences' is from William H. B. Aikins and F. C. Harrison, 'Recent observations on the therapeutic use of radium', *Canadian Practitioner and Review*, Vol. 36, pp. 1–10, 1911.

'The world has run raving mad on the subject of radium' is from George Bernard Shaw, *The Doctor's Dilemma: Preface on Doctors* (Constable, London, 1911), p. 12.

For the story of Eben Byers and his self-medication of Radithor, see Roger M. Macklis, 'The great radium scandal', *Scientific American*, Vol. 269, No. 2, pp. 94–99, August 1993.

The effects on workers of applying luminous paint is told in Claudia Clark, *Radium Girls: Women and Industrial Health Reform 1910–1935* (University of North Carolina Press, Chapel Hill, 1997). 'the best and happiest . . . actual progress' is from Marie Curie, *Pierre Curie: With Autobiographical Notes* (Macmillan, New York, 1923), pp. 186–187.

'Since the 1980s . . . health benefits of radon gas' is from Robbe Geysmans et al., 'Cure or carcinogen? A framing analysis of European radon spa websites', *International Journal of Public Health*, Vol. 67, p. 1, April 2022. The other two studies cited in the paragraph are Franke Annegret and Franke Thomas, 'Long-term benefits of radon spa therapy in rheumatic diseases: Results of the randomised, multi-centre IMuRa trial', *Rheumatology International*, Vol. 33, No. 11, pp. 2839–2850, July 2013; and Andreas Maier et al., 'Radon exposure: Therapeutic effects and cancer risk', *International Journal of Molecular Sciences*, Vol. 22, No. 1, 316, January 2021.

Sir William Ramsay's judgement that St Ives's pitchblende was 'very much richer' than that from Jáchymov is from *Times*, 20 October 1919. For the Radium Institute, see 'The Radium Institute', *The British Medical Journal*, Vol. 2, No. 2753, pp. 878–879, October 1913. St Ives's Trenwith mine and the town's water supply are detailed by Jim Hodge, as described by Rob Donovan in 'Under the surface at St Ives', 2013, http://www.robdonovan-author.co.uk/Under-Surface-StIves/UnderTheSurfaceOf-StIves.pdf.

For Johannes Mathesius, see Jan Royt, 'The mining town of Jáchymov: Reformation and art', Bohemian Reformation and Religious Practice, July 2003, http://www.brrp.org/proceedings/brrp5b/royt.pdf.

The story of J. Butler Burke's experiment with radium and sterilised bouillon and the sensational reaction to it and to his book *The Origin of Life, Its Physical Basis and Definition* is told in Campos, op. cit., pp. 56–99.

'All matter is alive . . . thesis' is in Campos, op. cit., p. 60. 'The biota and the rocks . . . buildings of universities' is from James Lovelock, 'The Earth as a living organism', in E. O. Wilson and F. M. Peter (eds.), *Biodiversity* (National Academies Press, Washington, 1988), Chapter 56, https://www.ncbi.nlm.nih.gov/books/NBK219276/.

'in my soul the decay . . . Science seemed to me destroyed' is from Hajo Düchting (ed.), *Wassily Kandinsky 1866–1944: A Revolution in Painting* (Taschen, Los Angeles, 2002), p. 10.

The exchange between Soddy and Rutherford about 'transmutation' is cited in Campos, op. cit., p. 13. Ramsay's statement, 'The philosopher's stone will have been discovered . . . the *elixir vitae*' is cited in Stephen Hilgartner et al., *Nuclear Language, Visions and Mindset* (Sierra Club Books, San Francisco, 1982), p. 4. The link between radium and the discovery of DNA is from Campos, op. cit., p. 266.

Soviet procurement of Jáchymov's uranium in the years after the Second World War is examined in Jiří Kašpárek, 'Soviet Russia and Czechoslovakia's uranium', *The Russian Review*, Vol. 11, No. 2, pp. 97–105, April 1952. For an account of the labour camps around Jáchymov, see Barbora Holá and Thijs Bouwknegt, '"Jáchymov's Hell": Trekking in the memoryscape of Czechoslovakia's Communist forced labour camps', *International Criminal Law Review*, Vol. 22, Nos. 1–2, pp. 1–19, October 2021, https://brill.com/view/journals/icla/22/1-2/article-p328_328.xml?language=en.

7. Aerolite

Sumerian, Hittite, Egyptian, Hebrew and Assyrian words for meteoric iron, as well as Aztec use of such iron, are discussed in T. A. Rickard, 'The use of meteoric iron', *The Journal of the Royal Anthropological Institute of Great Britain and Ireland*, Vol. 71, No. 1/2, pp. 55–66, 1941. For the Greenland meteorites and Peary's bringing them to New York, see P. A. M. Huntington, 'Robert E. Peary and the Cape York meteorites', *Polar Geography*, Vol. 26, No. 1, pp. 53–65, 2002.

For uses of meteoric iron in ancient Egypt, including burial objects in Tutankhamun's tomb, see Diane Johnson et al., 'Analysis of a prehistoric Egyptian iron bead with implications for the use and perception of meteorite iron in ancient Egypt', *Meteoritics & Planetary Science*, Vol. 48, No. 6, pp. 997–1006, May 2013.

For information on the Ka'aba stone, see Dr. Pervaiz Habibullah, '*Al-Hajar al-Aswad*' ('The Holy Black Stone'), Islamic Research

Foundation, undated, https://www.irfi.org/articles4/articles_5001
_6000 /Al-hajr%20aswad.htm

For the story of the holy aerolite at Emesa and Marcus Aurelius Antoninus, or Elagabalus, see P. M. Bellemare, 'Meteorite sparks a cult', *Journal of the Royal Astronomical Society of Canada*, Vol. 90, pp. 287–291, 1996.

Isaac Newton's translation of the Emerald Tablet, along with a number of other translations and a discussion of the tablet itself, can be found at https://sacred-texts.com/alc/emerald.htm. Newton's astonishing 'non-scientific' papers, including work on alchemy, are being made available online by the Newton Project, Faculty of History, University of Oxford, https://www.newtonproject.ox.ac. uk/about-us/newton-project.

For Naveen Jain's quote, 'Every single thing that we value on earth is in abundance in space', see 'More than just a rock collection: Meet Meteorite Man', 23 February 2016, https://www.youtube. com/watch?v=Eev0MDk2weY.

For the metal-rich asteroid, see Rachel Cormack, 'NASA's Hubble Telescope captures a rare metal asteroid worth 70,000 times the global economy', *Robb Report*, 28 October 2008, https://robbreport.com/lifestyle/news/rare-psyche-aster-oid-worth-way-more-than-the-global-economy-1234577976/?f-bclid=IwAR0UhBx9BlZ9im8SHzjEkdv-zA15ejjTkpg-Rt0AVAboGc0dmtFJ7qjkVcB4.

'may have been … the emergence of first life' is from Will Dunham, 'Asteroid discovery suggests ingredients for life on Earth came from space', *Reuters*, 23 March 2022, https://www.reuters.com/ lifestyle/science/asteroid-discovery-suggests-ingredients-life -earth-came-space-2023-03-21/.

For the meteor that may have formed the Prague Basin, see 'Czech impact crater detected', *Los Angeles Times*, 16 January 1989:

'A geologic footprint covering most of western Czechoslovakia may be the imprint of an immense object that smashed into Earth millions of years ago, scientists said last week. Based on a study of satellite images of central Europe taken from 22,500 miles in space, professors Michael Papagiannis and Farouk El-Baz said the shape is an apparent crater 200 miles in diameter. The researchers, both from Boston University, named the depression "Praha Basin" because the city of Praha (Prague) is near its center.'

For a general view of Rudolf's Prague, his court, his collecting and his interest in alchemy, I referred to Robert J. W. Evans, *Rudolf II and His World: A Study in Intellectual History 1576–1612* (Clarendon Press, Oxford, 1973); Eliška Fučíková et al., *Rudolf II and Prague: The Imperial Court and Residential City as the Cultural and Spiritual Heart of Central Europe* (Thames & Hudson, London, 1997); Vladimir Karpenko and Ivo Purš, *Alchemy and Rudolf II: Exploring the Secrets of Nature in Central Europe in the 16th and 17th Centuries* (Artefactum, Prague, 2016); and Peter Marshall, *The Mercurial Emperor: The Magic Circle of Rudolf II in Renaissance Prague* (Pimlico Press, London 2007), in which he describes Rudolf's collections as 'a veritable cornucopia of arcane phenomena . . . a systematic encyclopaedia of nature which could be used to unlock its secrets' (p. 77). A good view of Rudolf and his *Kunst- und Wunderkammer* can be found in Ivana Horacek, 'Alchemy of the gift: Things and material transformations at the court of Rudolf II', doctoral thesis submitted to the University of British Columbia, April 2015, https://open.library.ubc.ca/media/stream/pdf/24/1.0166243/1. For dimensions and other details of the *Kunst- und Wunderkammer*, see 'Holy Roman Emperor Rudolf II 1576–1612', Holy Roman Empire Association, undated, http://www.holyromanempireassociation.com/holy-roman-emperor-rudolf-ii.html. And for Rudolf and the interplay of the court and alchemy, see Pamela H. Smith,

'Alchemy as a language of mediation at the Habsburg Court', *Isis*, Vol. 85, No. 1, pp. 1–25, March 1994. For exploring the city, see Martin Stejskal, *Secret Prague* (Jonglez Publishing, Paris, 2018).

'What are the effects of fire . . . what is the whole force of hidden nature' is from Giambattista della Porta, *Magiae Naturalis* (1658), Book I, Chapter III. The full twenty-volume text can be found at https://web.archive.org/web/20180414134144/http://www.mindserpent.com/American_History/books/Porta/jportac1.html#Chap3Bk1.

'The Emperor is a lover of stones . . . all other things in creation', Anselmus de Boodt, is from his work *Gemma et Lapidum Historia*, published in 1609 and quoted in 'The Kunst- und Wunderkammer of Emperor Rudolf II', Die Welt der Habsburger, undated, https://www.habsburger.net/en/chapter/kunst-und-wunderkammer-emperor-rudolf-ii.

For 'the whole cosmos is present', see Karla Langedijk, 'The table in *pietre dure* for the Emperor. A new understanding of Rudolf II as a collector', *Mitteilungen des Kunsthistorischeschen Instituts in Florenz*, Vol. 42, Nos. 2/3, p. 374, 1998. For de Boodt's description of the table – 'mountains, rivers . . . inlaid invisibly', see ibid., p. 37; for the table as the 'eighth wonder of the world', see ibid., p. 358; and for the comparison with Ephesus, see ibid., p. 376.

'finest aviary in Europe' is from Marshall, op. cit., p. 53.

For moldavites, see 'Moldavite', *Geology Science*, 15 May 2023, https://geologyscience.com/minerals/moldavite/. See also Roman Skála et al., 'Moldavites from the Cheb Basin, Czech Republic', *Geochimica et Cosmochimica Acta*, Vol. 73, No. 4, pp. 1149–1179, February 2009. I am grateful too to the library of the Czech Geological Survey, Prague, and for conversations with Vít Kršul of the Moldavite Museum, Český Krumlov.

For mineral collectors, see 'Georgiana, Duchess of Devonshire', Buxton Museum, 17 October 2023, https://buxton-museumandartgallery.wordpress.com/2023/10/17/georgiana-duchess-of-devonshire/; and 'The Golden Age of collecting fine minerals', *The Arkenstone*, 13 September 2015, https://www.irocks.com/the-golden-age-of-collecting-fine-minerals.

For 'a powerful symbol of our connection . . . major spiritual growth and transformation' and 'it is not uncommon to see people faint or feel light headed due to its intense metaphysical energies,' see 'Moldavite – The celestial stone of transformation and awakening', *Eternal Spirit*, 8 March 2023, https://www.theeternalspirit.com.au/blogs/news/moldavite-the-stone-of-transformation. For 'this tumultuous tektite' and the TikTok craze, see Rebekah Harding, 'WitchTok is obsessed with this life-changing piece of glass', *Cosmopolitan*, 30 April 2021, https://www.cosmopolitan.com/lifestyle/a36267620/moldavite-tiktok-trend/.

For a detailed exploration of alchemy connections in Český Krumlov and the Rosenberg family and alchemy, see 'Alchemy in Český Krumlov', město Český Krumlov, undated, https://encyklo-pedie.ckrumlov.cz/en/mesto_histor_alchym/.

For the final stages of Rudolf's life, see Marshall, op. cit. pp. 223–225.

8. Mercury

For *Rasasiddhi* or 'knowledge of mercury', see Julie Sloane, 'Mercury: Element of the ancients', Dartmouth Toxic Metals, Dartmouth College Superfund Research Program, undated, https://sites.dart-mouth.edu/toxmetal/mercury/mercury-element-of-the-ancients.

The story of 'very significant' traces of mercury in Kepler's manu-scripts is from Alison Flood, 'Groundbreaking astronomer Kepler "may have practised alchemy"', *Guardian*, 13 June 2019, https://

www.theguardian.com/books/2019/jun/13/groundbreaking-astronomer-kepler-may-have-practised-alchemy. The hairs of the beard of his astronomy mentor Tycho Brahe also revealed very high levels of mercury; see Ludwig Jonas et al., 'Detection of mercury in the 411-year-old beard hairs of the astronomer Tycho Brahe by elemental analysis in electron microscopy', *National Library of Medicine*, Vol. 36, No. 5, pp. 312–319, October 2012.

For Theophrastus' ideas on mercury extraction, see Earle Caley and John Richards, *Theophrastus on Stones*, Graduate School Monographs, Ohio State University, No. 1, p. 196, 1956.

'The metals . . . smoky exhalation of the earth' by the semi-mythical Jabir ibn Hayyan is quoted in Lawrence M. Principe, *The Secrets of Alchemy* (University of Chicago Press, Chicago, 2012), p. 35.

'The paradox of *Du* . . . benefit and harm' is from Yan Liu, *Healing with Poisons: Potent Medicines in Medieval China* (University of Washington Press, Seattle, 2021), p. 30. For parallels between 'poison' and verbs meaning 'to give', see notes for Chapter 1, 'Ochre'.

Abraham Lincoln's adverse reaction to his mercury pills is explored and explained in 'UK lab reveals shocking mercury level in Lincoln's blue pills', Royal Society of Chemistry, 22 March 2010, https://www.rsc.org/news-events/articles/2010/03-march/blue-pills.

The story of Minamata's mercury contamination is told by Jun Ui, 'Minamata Disease', in Jun Ui (ed.), *Industrial Pollution in Japan* (United Nations University Press, Tokyo, 1992), pp. 103–132, https://archive.unu.edu/unupress/unupbooks/uu35ie/uu35ie00.htm.

The residual risk of mercury poisoning from former mining in Idrija is discussed in Alessandra R. G. Giumlia-Mair, 'The history of mercury production in the mine of Idrija, Slovenia', *Proceedings of the Second International Conference Archaeometallurgy in Europe,*

Aquileia, Italy, 2009, https://www.academia.edu/3612768/ The_History_of_Mercury_Production_in_the_Mine_of_ Idrija_Slovenia.

Biographical information about Paracelsus is sketchy, and for his travels he himself is the only source, so I have relied on the word 'claim' for the places he visited (and omitted some of the more fabulous ones). I have used Bruce T. Moran, *Paracelsus: An Alchemical Life* (Reaktion Books, London, 2019) and the preface of the English translation of his work: Arthur Edward Waite, *The Hermetic and Alchemical Writings of Philippus Aureolus Theophrastus Bombastus of Hohenheim, called Paracelsus the Great* (James Elliott, London, 1984). For Vol. I, see https://archive.org/details/ ArthurEWaiteHermeticAndAlchemicalWritingsOfParacelsus-VolI/page/n21/mode/2up, and for Vol. II, see https://archive. org/details/ArthurEWaiteHermeticAndAlchemicalWritingsOf-ParacelsusVolII.

'A doctor must seek out old wives, gypsies, sorcerers, wandering tribes ... we must hold fast that which is good' is quoted in Brendon Evans, 'Paracelsus – Father of toxicology, brother of general practice', *Australian Journal of General Practice*, Vol. 52, No. 6, p. 333, June 2023. Paracelsus's observation from Idrija – 'Everyone that lives here is bent ... being completely healthy again' is quoted in Matija Zorn et al., 'Creating new opportunities for an old mining region: The case of Idrija (Slovenia)', *Economic and Ecohistory*, Vol. 11, No. 1, pp. 123–138, 2015, https://core.ac.uk/download/ pdf/33289912.pdf. A curious observation of Paracelsus and his work was that, although his principal concern was the body and disease, he avoided writing about sex. The story emerged that he had been castrated (by a pig on a dunghill, said some). His physical remains are kept in a church in Salzburg and show an unusually broad pelvis. 'Forensic specialists suggest that Paracelsus was either a genetic male afflicted with pseudohermaphroditism or a genetic female suffering from adrenogenital syndrome . . . we are left with the remarkable possibility that the gender of Paracelsus may have been capable of description as either female or male.' From William R. Newman, *Promethean Ambitions: Alchemy and the Quest to Perfect Nature* (University of Chicago Press, Chicago, 2004), p. 197.

'All things are poison . . . not a poison' is quoted and discussed in W. B. Deichmann et al., 'What is there that is not poison? A study of the Third Defense by Paracelsus', *Archives of Toxicology*, Vol. 58, No. 4, pp. 207–213, April 1986.

'There is nothing in Heaven . . . is not in man' is quoted in Joseph F. Borzelleca, 'Paracelsus: Herald of modern toxicology', *Toxicological Sciences*, Vol. 53, No. 1, pp. 2–4, January 2000. 'He who is born in imagination . . . a new heaven' is quoted in 'Paracelsus', *Britannica*, 16 May 2024, https://www.britannica.com/biography/Paracelsus. 'Where is God? Listen, you blind human, you live in God and God in you' is from Jakob Böhme '*Morgenröte im Aufgang*' ('Aurora'), in Will-Erich Peukert (ed.), *Jakob Böhme Sämtliche Schriften* (Friedrich Frommann Verlag, Stuttgart, 1955), Vol. 1, p. 327. 'The imagination is not a state: it is the human existence itself' is from William Blake, *Milton, Book the Second*. 'Imagination is the real and eternal world of which this vegetable universe is but a faint shadow' is from William Blake, *Jerusalem: The Emanation of the Giant Albion*, Chapter I, Plate 77. 'microcosm, a condensation of the entire universe' is quoted in Bruce T. Moran, 'Chemical medicine, theory', Reading Early Medicine, Johns Hopkins University, 21 September 2022, https://readingearlymedicine.org/initiatives/chemical-medicine-theory/.

For biographical information on Scopoli, see Alfred B. Kobal and Darja K. Grum, 'Scopoli's work in the field of mercurialism in light of today's knowledge: Past and present perspectives', *American Journal of Independent Medicine*, Vol. 53, No. 5, pp. 535–547, May 2010;

and Darinka Soban, *Joannes A. Scopoli – Carl Linnaeus; Dopisovanje / Correspondence 1760–1775* (Prirodoslovno Drustvo, Ljubljana, 2004), in which all the letters between Scopoli and Linnaeus are translated. For Linnaeus's classification of minerals, see Rachel Laudan, *From Mineralogy to Geology: The Foundations of a Science, 1650–1830* (University of Chicago Press, Chicago, 1987), p. 285.

'There are as many mercuries as there are things' is from 'Paracelsus (1493–1541)', *Encyclopedia.com*, 21 May 2018, https://www.encyclopedia.com/people/medicine/medicine-biographies/philippus-aureolus-paracelsus.

The heavily symbolic image of Mercury is from the *Rosarium Philosophorum* and reproduced and discussed in Vladimir Karpenko and Ivo Purš, *Alchemy and Rudolf II: Exploring the Secrets of Nature in Central Europe in the 16th and 17th Centuries* (Artefactum, Prague, 2016), pp. 55–59.

'these myths are emphatic . . . most tragic and most debased' is from Mircea Eliade, *The Forge and the Crucible: The Origins and Structures of Alchemy*, 2nd ed. (University of Chicago Press, Chicago, 1978), p. 67.

For the works of Hermes Trismegistus, see *The Corpus Hermeticum: Initiation into Hermetics, the Hermetica of Hermes Trismegistus*, trans. G. R. S. Mead (Pantianos Classics, 1906); and *The Divine Pymander*, trans. John Everard (Robert White, London, 1650).

9. Copper

The story of the copper ingot and de Champlain is in T. A. Rickard, *Man and Metals: A History of Mining in Relation to the Development of Civilisation* (McGraw-Hill, London, 1932), p. 102.

For Julius Eldred's efforts to transport the Ontonagon Boulder, and its subsequent travels, see Gustave Lester, '"Native copper": Exhibiting Anishinaabe wealth at the US National Museum', *Early American Studies Miscellany*, 20 May 2024, https://web.sas.upenn.edu/earlyamericanstudies/2024/05/20/native-copper-exhibiting-anishinaabe-wealth-at-the-u-s-national-museum-gustave-lester/.

For an introduction to the archaeological work at Mitterberg, see Ernst Pernicka et al., 'Bronze Age copper produced at Mitterberg, Austria, and its distribution', *Archaeologia Austriaca*, Band 100/2016, pp. 19–55, 2016; and E. Breitenlechner et al., 'An interdisciplinary study on the environmental reflection of prehistoric mining activities at the Mitterberg Main Lode', *Archaeometry*, Vol. 56, No. 1, pp. 102–128, March 2013.

On the subject of Bronze Age deposition, I am grateful to Dr Matt Knight (Senior Curator of Early Prehistory at the National Museum of Scotland) for a private discussion. See also his lecture at the Scottish Society of Antiquaries, 'Fragments of the Bronze Age', 9 March 2020, https://www.youtube.com/watch?v=zO7U5E5EY6g.

'to support and enhance . . . properly functioning' is from Mary W. Helms, 'Nourishing a structured world with living metal in Bronze Age Europe', *World Art*, Vol. 2, No. 1, p. 106, July 2012.

For Isaac Casaubon's debunking of Hermes Trismegistus, see Anthony Grafton, 'Protestant versus prophet: Isaac Casaubon on Hermes Trismegistus', *Journal of the Warburg and Courtauld Institutes*, Vol. 46, No. 1, pp. 78–93, 1983.

Until it was purchased by John Maynard Keynes at a Sotheby's sale in 1936, Newton's work on alchemy and esoteric matters was largely unknown; Newton himself kept it quiet for fear of ridicule and punishment – some alchemical practice was banned during his lifetime with the ultimate sanction of public hanging from a gilded gallows. Newton as 'the last of the great magicians' was part of a lecture written by John Maynard Keynes and delivered posthumously by his brother, Geoffrey Keynes, who was a Blake scholar and produced one of the principal editions of Blake's work. The lecture was entitled 'Newton the Man', and the text can be found at https://mathshistory.st-andrews.ac.uk/Extras/Keynes_Newton/.

'no boundary between the cosmos, humanity, and earth' and 'pattern thinkers . . . from doing things' are both from Christopher Bamford, 'Green Hermeticism: David Levi Strauss with Peter Lamborn Wilson & Christopher Bamford', *Brooklyn Rail*, December 2007/January 2008, https://brooklynrail.org/special/RIVER_ RAIL/river-rail/David-Levi-Strauss-with-Peter-Lamborn-Wilson. See also Peter Lamborn Wilson et al., *Green Hermeticism: Alchemy and Ecology* (Lindisfarne Books, Great Barrington, 2007).

For William Blake, I used a variety of sources. To read his works, I reached for my own copy of Geoffrey Keynes (ed.), *Poetry and Prose of William Blake* (Nonesuch, London, 1927). But most references to Blake's work use David V. Erdman (ed.), *The Complete Poetry and Prose of William Blake* (Doubleday, New York, 1988), so I have sourced my quotations from Blake mainly from this edition. Also by my side for these sections were Richard Holmes (ed.), *Gilchrist on Blake: The Life of William Blake by Alexander Gilchrist* (Harper Perennial, New York, 2011), which was first published in 1863; Peter Ackroyd, *Blake* (Vintage, London, 1999); and John Higgs, *William Blake vs the World* (Weidenfeld & Nicolson, London, 2021).

'Bacon & Newton . . . / Water-wheels of Newton' is from William Blake, *Jerusalem: The Emanation of the Great Albion* (composed 1804–c.1820), Plate 54, in Erdman, op. cit., p. 159. 'Los with his mace of iron / . . . breaks the potsherds' is from *Jerusalem*, Plate 78, Erdman, op. cit., p. 233. 'All things acted on Earth / . . . from these Works' is from *Jerusalem*, Plate 16, Erdman, op. cit., p. 161.

Blake's techniques of etching, engraving and print-making were innovative and wide-ranging. I am grateful for a private discussion with the print-maker Michael Phillips (who is also a leading expert on Blake's techniques), correspondence with Blake scholar Professor Jason Whittaker and a visit to watch Brian Hanscomb at work in his studio on the edge of Bodmin Moor; Hanscomb is rare among print-makers in engraving in copper rather than etching. I also consulted Robert Essick, *William Blake, Printmaker* (Princeton University Press, Princeton, 1980); and G. E. Bentley, Jr, 'William Blake's techniques of engraving and printing', *Studies in Bibliography*, Vol. 34, pp. 241–253, 1981.

'never suspended his Labours on Copper for a single day' was part of Blake's *Chaucer Prospectus, Second, Composite Draft; see* Erdman, op. cit., p. 568.

'mechanical Excellence is the Only Vehicle of Genius' is from William Blake, *Annotations to The Works of Sir Joshua Reynolds* (London, 1798), ed. Edmond Malone, p. 14.

'the notion that man has a body . . . which was hid' is from William Blake, *The Marriage of Heaven and Hell*, Plate 14, Erdman, op. cit., p. 39. There is a thorough discussion of the method in Michael Phillips, '"Printing in the infernal method": William Blake's method of "illuminated printing"', *Interfaces*, Vol. 39, pp. 67–89, July 2018.

The Thames from Blake's window looking 'like a bar of gold' was a description by George Richmond, one of 'the Ancients', a group of young men, including Samuel Palmer, who visited and revered Blake in his last years, quoted in Angus Whitehead, 'William Blake's last residence: No. 3 Fountain Court, Strand. George Richmond's plan and an unrecorded letter to John Linnell', *The British Art Journal*, Vol. 6, No. 1, p. 26, Spring/Summer 2005. Fountain Court was in a space near the Savoy Steps, the alley where in 1965 Bob Dylan recorded his cue-card video for 'Subterranean Homesick Blues'.

'glaring and flickering light, Blake is greater than Rembrandt' is from John Ruskin, *The Elements of Drawing in Three Letters to Beginners* (Smith, Elder & Co., London, 1857), p. 194.

'to find a more beautiful chapter . . . these Job designs' is from Joseph Wicksteed, *Blake's Vision of the Book of Job – A Study* (Dent & Sons, London, 1924), p. 34.

For the story of the Trianon Press, see Zoë Ingalls, 'The Trianon Press: a "triumph of enthusiasm over reason"', *The Chronicle of Higher Education*, 17 July 1991, https://www.chronicle.com/article/the-trianon-press-a-triumph-of-enthusiasm-over-reason/; and for Arnold Fawcus, see Geoffrey Keynes, 'Arnold Fawcus', *Blake/An Illustrated Quarterly*, Vol. 13, No. 2, pp. 110–111, Fall 1979, http://bq.blakearchive.org/pdfs/13.2.keynes.pdf.

For general context to the diffusion of knowledge during the Renaissance, and some of its more colourful characters, I recommend Anthony Hobson, *Humanists and Bookbinders: The Origins and Diffusion of the Humanistic Bookbinding, 1459–1559* (Cambridge University Press, Cambridge, 1989).

Blake's series *Illustrations to the Book of Job* was closely scrutinised during the twentieth century, with a wide range of Freudian, biblical and imaginative interpretations. 'An enduring aspect of the Biblical story of Job is the constant temptation on the part of the reader to redefine and reinterpret it in the light of personal experience and belief' could be equally well applied to Blake's own interpretation, and the interpretations of his work; see Alison Sanders McFarland, 'A deconstruction of William Blake's vision: Vaughan Williams and Job', *International Journal of Musicology*, Vol. 3, pp. 339–371, 1994. As well as Wicksteed's *Blake's Vision*, I consulted Andrew Wright, *Blake's Job: A Commentary* (Oxford University Press, Oxford, 1972); Milton Percival, *William Blake's Circle of Destiny* (Octagon Books, New York, 1964); and the more recent Morton Paley, *The*

Traveller in the Evening: The Last Works of William Blake (Oxford University Press, New York, 2003). An example of how the Job series can be used to illustrate a psychological and creative state can be seen in Jussi Antti Saarinen, 'The concept of the oceanic feeling in artistic creativity and in the analysis of visual artworks', *The Journal of Aesthetic Education* , Vol. 49, No. 3, pp. 15–31, Fall 2015. My own reading of the series (and much of Blake's work) as being proto-ecological – sharing the same timeless principles of universality, transcendence, interconnectedness and the union of the physical and the spiritual – was fed by various streams. Jonathan Bate, *Romantic Ecology: Wordsworth and the Environmental Tradition* (Routledge, London, 1991) introduces the subject, and it is picked up in Mark S. Lussier, 'Blake's deep ecology', *Studies in Romanticism*, Vol. 35, No. 3, Green Romanticism, pp. 393–408, Fall 1996; and Kevin Hutchings, *Imagining Nature: Blake's Environmental Poetics* (McGill-Queen's University Press, Montreal, 2003).

'Blake touches the height of his genius . . . eternal rapture' is from Wicksteed, op. cit., p. 102.

Die blaue Blume, 'the blue flower', became a popular symbol for the romantic quest, a striving for the unattainable, following the publication of Novalis's unfinished novel *Heinrich von Ofterdingen* in 1802. Some have linked Goethe's search for the *Urpflanze*, 'the plant archetype', as an earlier model for Novalis. Novalis draws on a late-medieval story for *Heinrich von Ofterdingen*. His novel was first translated into English by Friedrich von Schlegel and Ludwig Tieck (John Owen, Cambridge, 1842).

'if he be still living' is in a letter from Charles Lamb to Bernard Barton, 15 May 1824, from *Works of Charles and Mary Lamb. VI–VII. Letters*, https://www.lordbyron.org/monograph.php?doc=-ChLamb.1905&select=L1824.8.

'a Memento in time to come & to speak to future generations by a Sublime Allegory' is from a letter from William Blake to Thomas Butts, 6 July 1803, Erdman, op. cit., p. 729. The comment about the Job prints being 'too much Finished, or over Labour'd' is from a letter from William Blake to John Linnell, 15 March 1827 citing George Cumberland's difficulties in selling the prints to his Bristol friends; see Erdman, op. cit., p. 782.

'in which the images of children . . . human existence' is from Ackroyd, op. cit., p. 388; and 'had been very near the gates of death . . . The Imagination which Liveth for Ever' is quoted in op. cit., pp. 388–389. For an account of Blake's final hours, see Holmes, op. cit., pp. 379–382.

For the possible link between copper poisoning and Blake's death, see Lane Robson and Joseph Viscomi, 'Blake's Death', *Blake/An Illustrated Quarterly*, Vol. 30, No. 2, Fall 1996, https:// bq.blakearchive.org/30.2.robson. In the entry for William Blake in *Oxford Dictionary of National Biography*, Robert Essick writes: 'The precise cause of Blake's death is not known, but the most detailed analysis of the symptoms suggests biliary cirrhosis, possibly caused by years of inhaling cupreous fumes while etching.' For the medical context of such poisoning, see M. Worwood et al., 'Copper and manganese concentrations in biliary cirrhosis of liver', *The British Medical Journal*, Vol. 3, No. 5614, pp. 344–346, August 1968, https://www.jstor.org/stable/20393618?seq=1.

10. Gold

The Inca belief that gold is the sweat of the sun god is from Neil MacGregor, 'Inca gold llama' from *History of the World in 100 Objects*, BBC/British Museum, 11 August 2021, https://www.bbc. co.uk/programmes/b00tn9vj. For the Aztec belief that gold was the excrement of the sun god, see Cecilia F. Klein, 'Teocuitlatl, "divine excrement": The significance of "holy shit" in ancient Mexico', *Art Journal*, Vol. 52, No. 3, pp. 20–27, September 1993. (Silver was *iuac teocuitlatl*, 'white sacred excrement'.)

For the necropolis at Varna in Bulgaria, see Marija Gimbutas, 'Gold treasure at Varna', *Archaeology*, Vol. 30, No. 1, pp. 44–51, January 1977.

'Gold is the source of crime . . . maddened against brothers' is quoted in Georgius Agricola's *De Re Metallica*, op. cit., p. 7.

For details of the temple at Coricancha, see Jamake Highwater, 'Rediscovering the Inca heritage: Temple of the sun', *Archaeology*, Vol. 38, No. 2, pp. 16–21, March/April 1985; and Joanne Pillsbury, 'Gold in the ancient Americas', The Metropolitan Museum of Art, July 2020, http://www.metmuseum.org/toah/hd/gdaa/hd_gdaa.htm.

The story of John Deason's gold nugget is told by Michael G., 'Welcome Stranger Gold Nugget Anniversary', *Georarities*, 8 February 2021, https://georarities.com/2021/02/08/welcome-stranger-gold-nugget-anniversary/.

The account of Las Médulas gold mine and its light-starved miners is in Pliny the Elder, *Naturalis Historia*, XXXIII, 70; and 'the miners gaze as conquerors on the collapse of nature' is from ibid., XXXIII, 73.

'Gold is immortality' and the ritual description are from *Shatapatha Brahmana*, Kanda VII, adhyaya 4, brahmana 10–15.

The Queen of Sheba's gifts to King Solomon is from 1 Kings 10:10.

The full name of the Asante stool is Sika Dwa Kofi, 'Golden Stool born on a Friday'. 'The royal stool . . . of solid gold' is quoted in 'Asante Gold', Victoria & Albert Museum, undated, http://www.vam.ac.uk/content/articles/a/asante-gold/: 'In 1817 the British envoy Thomas Bowdich spent three months in Kumasi where he received a state reception. He left a vivid description of the richly ornate court of the Asantehene Osei Bonsu.' The quote from Lloyd George is from 'The Ashanti expedition', *Hansard*, UK Parliament, 19 March 1901, https://hansard.parliament. uk/Commons/1901-03-19/debates/19ea3bcc-b7f4-4dab-9afc-0b9abe700ab4/TheAshantiExpedition.

The adoption of the gold standard by Isaac Newton, and how it lasted and was eventually abandoned, is told by Michael D. Bordo, 'Gold standard', *The Library of Economics and Liberty*, undated, https://www.econlib.org/library/Enc/GoldStandard.html.

'Everything is groaning . . . sounds louder still' is from Maxim Gorky, *City of the Yellow Devil – Pamphlets, Articles and Letters about America* (Progress Publishers, Moscow, 1972), p. 10.

'ten million men and maimed another thirty million' and 'When we are victorious . . . public lavatories' are from V. I. Lenin, 'The importance of gold now and after the complete victory of socialism', *Pravda*, No. 251, 6–7 November 1921, in *Lenin's Collected Works*, 2nd English ed., trans. David Skvirsky and George Hanna (Progress Publishers, Moscow, 1965), Vol. 33, pp. 109–116. For the Bolshevik attitude to gold, I also used Veronica Davidov, 'Soviet gold as sign and value: Anthropological musings on literary texts as cultural artifacts', *Etnofoor*, Gold, Vol. 25, No. 1, pp. 14–28, 2013.

The story of *Torgsin* and the Soviet need for gold is told in Elena Osokina, *Stalin and Stalin's Quest for Gold* (Cornell University Press, New York, 2021), https://www.cornellpress.cornell.edu/torgsin-and-stalins-quest-for-gold/.

For the archaeological context at Sakdrisi and an account of the excavations and finds, I consulted Thomas Stöllner and Irina Gambashidze (eds.), *The Gold of Sakdrisi: Man's First Gold Mining Enterprise* (VML Verlag, Rahden, 2016); and Thomas Stöllner and Irina Gambashidze, 'The gold mine of Sakdrisi and the earliest mining and metallurgy in the Transcaucasus and the Kura-Valley system', *Proceedings of International Conference: Problems of Early Metal Age Archaeology of Caucasus and Anatolia*, Tbilisi, Georgia, 19–23 November 2014, pp. 102–125. For the campaign against mining at Sakdrisi, see Rayhan Demytrie, 'Georgia's gold mine dilemma', BBC News, 29 May 2014, https://www.bbc.co.uk/news/world-europe-27499882.

For the geology of Georgia, with particular reference to gold, I drew on Sergo A. Kekelia et al., 'Gold deposits and occurrences of the Greater Caucasus, Georgia Republic: Their genesis and prospecting criteria', *Ore Geology Reviews*, Vol. 34, No. 3, pp. 369–386, November 2008; and Geological Institute of Academy of Sciences of Georgia, Caucasian Institute of Mineral Resources, U.S. Geological Survey, https://www.sciencedirect.com/science/article/pii/S0169136808000309?via%3Dihub

For gold in the region of Colchis, see Avtandil Okrostsvaridze and David Bluashvili, 'Mythical "gold sands" of Svaneti (Greater Caucasus, Georgia): Geological reality and gold mining artefacts', *Bulletin of the Georgian National Academy of Sciences*, Vol. 4, No. 2, pp. 117–121, February 2010.

Strabo alludes to the Svan practice of extracting gold in Strabo, *The Geography of Strabo*, Loeb Classics (Harvard University Press, Cambridge, Massachusetts, 1928), Book IX, Chapter 2, p. 217: 'It is said that in their country gold is carried down by the mountain-torrents, and that the barbarians obtain it by means of perforated troughs and fleecy skins, and that this is the origin of the myth of the golden fleece.'

For a historical overview of gold-mining, see Spyridon Mathio-udakis et al., 'Alluvial gold mining from Argonauts to Agricola', *Material Proceedings*, Vol. 15, No. 1, October 2023.

For 'the most prestigious and historic order of chivalry in the world' and an overview of the chivalric order, see J. Paul Murdock, 'Distinguished Order of the Golden Fleece', *A Royal Heraldry*, 5 March 2021, https://aroyalheraldry.weebly.com/blog/orders-of-chivalry-spain.

The connection between Jason and alchemy is considered by Jason Colavito, 'Jason and alchemy 17th century', Jason and the Argonauts Through the Ages, 2014, http://www.argonauts-book.com/jason-and-alchemy.html.

An account of a banquet for the Order of the Golden Fleece, in which the order's founder Philip the Good, Duke of Burgundy, swore an oath on a living pheasant to recapture the city of Byzantium, is told by Edward Gibbon, *Decline and Fall of the Roman Empire*, ed. Felipe Fernández-Armesto (Folio Society, London, 1990), Vol. 8, p. 246.

For Mount Mezir, see Richard Baerug, *Svaneti, The Essence of the Caucasus* (Ushba Press, Georgia, 2019), p. 101.

Gold-mining as 'one of the most destructive industries in the world' is from 'Environmental impacts of gold mining', Earthworks, undated, https://earthworks.org/issues/environmental-impacts-of-gold-mining/. Earthworks is a group engaged in 'preventing the destructive impacts of the extraction of oil, gas and minerals'. See also Louisa J. Esdaile and Justin M. Chalker, 'The mercury problem in artisanal and small-scale gold mining', *Chemistry – A European Journal*, Vol. 24, No. 27, pp. 6905–6916, May 2018.

For the comment 'stunning absurdity' on *Mackenna's Gold*, see Vincent Canby, 'The screen: "Mackenna's Gold" in Apache Country', *New York Times*, 19 June 1969. For Gregory Peck's 'wretched' dismissal, and other details of the film, see Moira Finnie, 'MacKenna's Gold (1969): A fool and his money', *Silver Screen Oasis*, 18 July 2011, https://www.silverscreenoasis.com/oasis3/viewtopic.php?t=5313.

'The bottom of the world is gold . . . upside down' is from Jack Kerouac, *On the Road* (Viking Press, New York, 1959), p. 158.

11. Lithium

For lithium's unusual chemical behaviour under pressure, see Stephanie A. Mack et al., 'Emergence of topological electronic phases in elemental lithium under pressure', *Proceedings of the National Academy of Sciences of the United States of America*, Vol. 116, No. 19, pp. 9197–9201, March 2019.

The health benefits of lithium, including Lithia Springs, 7Up, and President Cleveland's call, are outlined in Walter A. Brown, *Lithium: A Doctor, a Drug, and a Breakthrough* (Liveright, New York, 2019), pp. 57–59. For lithium as a 'gold standard' treatment for bipolar disorder, see ibid., pp. i–xvi.

For the Salar de Uyuni lithium deposits in Bolivia, see Clifford Krauss, 'Green-energy race draws an American underdog to Bolivia's lithium', *New York Times*, 16 December 2021; and for the environmental hazards of the rush for lithium, see Oliver Balch, 'The curse of "white oil": Electric vehicles' dirty secret', *Guardian*, 8 December 2020.

Soil

The figure of a million earthworms per acre is from https://worldagriculturesolutions.com/2016/05/11/earthworm-primer/

For arthropod abundance, see Diksha Tokas et al., 'Plant-microbe interactions: Role in sustainable agriculture and food security in a changing climate', in *Plant-Microbe Interaction - Recent Advances in Molecular and Biochemical Approaches*, Vol. 2 (Academic Press, Cambridge, Massachusetts, 2023), pp. 363–391, https://www.sciencedirect.com/science/article/abs/pii/B9780323918763000087.

For nematodes making up 80% of all animals, see Johan van den Hoogen et al., 'Soil nematode abundance and functional group composition at a global scale', *Nature*, Vol. 572, pp. 194–198, July 2019.

For protozoa, see F. Ekelund and R. Rønn, 'Notes on protozoa in agricultural soil with emphasis on heterotrophic flagellates and naked amoebae and their ecology', *FEMS Microbiology Reviews*, Vol. 15, No. 4, pp. 321–353, December 1994.

For bacteria in soil, see Julie Bobyock, 'Improving climate predictions by unlocking the secrets of soil microbes', *Berkeley Lab*, 5 February 2024, https://newscenter.lbl.gov/2024/02/05/improving-climate-predictions-by-unlocking-the-secrets-of-soil-microbes/.

For 'eight miles of mycelial fibres', see Paul Stamets, *Mycelium Running: How Mushrooms Can Help Save the World* (Ten Speed Press, Berkeley, 2005), quoted in Bill Keep, 'Fungi are fascinating', undated, https://ucanr.edu/sites/Shasta_College_Master_Gardener/files/169405.pdf.

ACKNOWLEDGEMENTS

This book began a long time ago, and only much later became a journey. I am indebted to many people who have helped along the way. As a child rock collector, I was encouraged by a large number of adults, but none more so than an elderly woman known to me only as Miss Wynne Willson, whom I struggled to keep up with on stone-hunting expeditions around the West Country's cliffs, quarries and dual carriageway works. I am grateful more recently to all at the Royal Cornwall Museum with its world-class Rashleigh mineral collection – to the curators Sara Chambers, Jayne Wackett, Jeni Woolcock; to the late Angela Broome, who presided over the Courtney Library in Truro (like many researchers, I found that I only had to mention a Cornish subject or place or person for my desk to fill with piles of relevant material); to Sam Perkin and Patrick Moret and other members of the Carbis Bay Crew for sharing their love of ropework and the tighter spots of Cornwall's mine-riddled underworld; to archaeologists Nicholas Johnson and Pete Herring, whose knowledge of the distant and not-so-distant past is matched only by their enthusiasm for it; to Professor Frances Wall of the Royal Cornwall Geological Society and to Daniela Recht, who gamely assisted in my efforts to learn German.

In the Netherlands, I owe my thanks to Jippe Kreuning, Wim and Helen van Wijk and Rob van de Made; on the Rhine to Franz and Sonja Schramm and the *SS Seestern*; to Courtenay Smale, Professor Gregor Borg and Dr Regine Maraszek for filling in the story of the Nebra Sky Disc; in Stolberg on the silver trail, thanks to Jost and Sylviane zu Stolberg-Stolberg, Christoph zu Stolberg; in Prague to Benedict Allen, Adam von Pezold, Vladimir Karpenko; in southern Bohemia, Vít Kršul; in Austria, Plum Webber, Elizabeth von Pezold and Heinrich von Pezold; and in Slovenia, Bojan Režun, Mojca Gorjup Kavčič, Marija Terpin, Tatjana Dizdarevic and all at UNESCO Geopark Idrija; in Bosnia, Bojana Mojsov and Julian Reilly; in Bochum and also in the Austrian Alps, Professor Thomas Stöllner, Dr Peter Thomas, Dr Jennifer Garner and the team from the Deutsches Bergbau-Museum, Department of Mining Archaeology (*Montanarchäologie*); in Georgia, Dr Irina Gambashidze, Mikheil Abramishvili, Lasha Orjonikidze, Narkis Nigoriani, Eka Pilpani, Caroline Eden and Richard Baerug.

Many thanks to Chris Harker, Jeremy Wrathall and Lucy Crane at Cornish Lithium, to Professor Jason Whittaker, Michael Phillips and Brian Hanscomb. Thanks too to Clem Cecil, Martin and Miranda Thomas, Sally and Christopher Tennant, Hugo Tagholm and metalworker and knife forger Ronen Burstein.

Anna Webber, as always, proved the best of agents and as an Austrian was able to offer ideas and a subtle understanding of Mitteleuropa. As editor, Laura Barber has an alchemist's ability to spot the intentions that often lie hidden inside a manuscript and draw them out. Jack Alexander's forensic editorial input was hugely impressive, likewise that of Kate Shearman. Also to Pru Rowlandson, Christine Lo and all at Granta.

To Clio and Arthur, who contributed to these pages in ways that are both oblique and precious; and last, to Charlotte for patient kitchen-table editing, critical discussion and support in a hundred guises, this book is dedicated to you, with love.

INDEX

Achilles, 69
aerolites, 153, 167–8, 170
agate, 1, 183
Aikins, William, 130
Akhmatova, Anna, 94
alchemy, 19–20, 113, 116,
    146, 148, 155, 159–
    60, 165–6, 174–5,
    194, 202, 204–7, 212,
    221–6, 249, 251, 258
    Blake and, 224–6
    and mercury, 184, 186,
    204–6
    Newton and, 223–4
    and procreation, 205–6
    see also Hermetic
    tradition; natural
    magic; Philosopher's
    Stone; transmutation
Alcott, Louisa May, 187
Alexander I, Tsar, 171
Alexandria, 184, 209–10,
    223
alexandrite, 171
'alkahest', 211
Altamira Cave, 7
Altenau, 109–11, 116
Alzheimer's disease, 215,
    278
American Museum of
    Natural History, 128,
    152
American Radium Society,
    130

amethyst, 170–1
amino acids, 155
ammonites, 2, 4, 91, 289
Amsterdam, 44, 48, 50
Anning, Mary, 177
aquamarines, 170
Arctic ice cores, 108
arenites, 158
argon, 147
arsenic, 8, 16–18, 69, 138,
    218, 286
arsenopyrite, 171
Asante Empire, 250
astrology, 164
astronomy, 164–5, 284
Asur, 69, 208
atomic theory, 147
Aurum Potabile, 166
Australia, 7–8
australopithecines, 170
Aymara, 278
Ayurvedic medicine, 17
Aztecs, 154, 246
azurite, 8, 213

Babylonians, 68
Bacon, Francis, 222, 224
Baghdad, Abbasid, 20
bal maidens, 208
balneotherapy, 142
Bamford, Christopher,
    223–4
batholiths, 11
bauxite, 84

Bear constellation, 70
Beardy Weirdies, 20–1, 27
beavers, 65
Behounek spring, 140
belemnites, 1
Bell, Alexander Graham,
    130
Beltane, 93
Benjamin, Walter, 238
Berbero-Egyptian
    mythology, 154
Berbers, 249
Berdyaev, Nikolai, 60–1
Berlin, Brent, 6
Berlin Wall, collapse of, 120
Biesbosch, 51–2, 54, 61–2,
    64, 73, 217
Biokeet, 51, 55, 61, 66
bipolar disorder, 278
Black Stone of Mecca, 153
Blake, William, 115, 196,
    224–33, 242–5, 285–7
Blok, Alexander, 94
Blombos Cave, 6
bloodstone, 162
bludovite, 171
Blue John, 1
Bochum, 217, 222, 236,
    240–1, 254–5
Bodleian Library, 230
Bodmin Moor, 11, 20, 261
Bodwen, 26
Bogomils, 4
Bohemia, 99, 132, 156, 159,

161–3, 165–6, 168, 173–6, 178, 182–3, 238
Böhme, Jakob, 196, 211
Book of Genesis, 146, 233
Borg, Gregor, 87–9
Borgias, 17
Borkenkäfer beetle, 108–9
Bouffadou fire pipes, 205
Boyle, Robert, 222
Brahe, Tycho, 164, 175
brass doorknobs, 215
Briggs, Raymond, 136
Brighid, 41
British Museum, 90
Brochocka, Helena, 135
Brocken, 93, 113, 116–18, 120
Brontë, Charlotte, 187
bronze, 15–16, 22, 25, 67–72, 78–81, 86–93, 148, 154, 213, 219–20, 240
see also Nebra Sky Disc
Bronze Age, 11–12, 15, 22, 25, 40, 42, 45, 72, 74, 78–80, 90, 132, 213, 216–17, 219–20, 222, 236–7, 240, 241, 254–5, 280
Bruno, Giordano, 164
Burke, J. Butler, 146–8
Burns, Robert, 187, 269
Buryat, 69
Byers, Eben, 131, 148
Byzantium, 46, 211, 258

cadmium, 286
Caerhays Castle, 89
California gold rush, 251
camels, 248–9
Campos, Luis A., 128, 148
cantarella, 17
carbon dioxide, 43–4, 125
carnelian, 80
Carnmenellis, 10, 12, 16, 32, 45, 87, 89, 280

Carnon, river and valley, 12, 87, 89
Cartesian dualism, 226
Casaubon, Isaac, 222
Cash, Johnny, 234
Cashel Man, 43
Cassiterides, 25
Catalan Atlas, 250
Cathars, 4
Catherine the Great, 171
Catholic Church, 98
Cerro Rico de Potosí, 94–9, 124, 204, 279
cerussite, 170
Cervantes, Miguel de, 94
Český Krumlov, 167–9, 172, 175–6, 182–3
chalcedony, 162
Chalcolithic Age, 213
chalcopyrite, 218, 237
Champlain, Samuel de, 214
Channel Tunnel, 21
Charter of the Stannaries, 30
Chaucer, Geoffrey, 91
Chauvet Cave, 7
Chernobyl nuclear disaster, 144
Cherokee, 278
Chewa, 208
China
  alchemy, 20, 155, 186, 249
  Gun-Yu flood, 57
  origin of chicken, 80
  silver trade, 95–6, 98–9
  Terracotta Army, 186
  and Svaneti gold mining, 262–3, 265–6
  Xia and Shang dynasties, 68
Chinese medicine, 16
cinnabar, 188, 208
citrine crystals, 170
Clausthal, 15, 106
Cleopatra the Alchemist, 184

Cleveland, Grover, 278
Clive, Lady, 170
cobalt, 24, 138, 170, 285
Cocteau, Jean, 185
coinage, debasement of, 98
Colchis, 257–8, 262
Colorado beetle, 18
conflict diamonds, 178
conflict minerals, 24
Conquistadores, 248
Copernicus, Nicolaus, 98
copper, 8, 10–14, 16, 22, 24, 45, 49, 84, 91, 96, 141, 143, 152, 154, 170, 213–45, 247, 251, 255, 257, 281–2
  'blue mine', 237–40
  and bronze, 22, 25, 67, 69, 72, 213
  colours, 213
  conductivity, 215
  deficiencies, 214–15
  enters food chain, 285
  and Nebra Sky Disc, 87–9, 217, 220
  and relief etching, 245
  speleothems, 37–8
  smelting, 220–1
  toxicity, 215
copperplate printing, 215–16, 225–8, 242, 244–5
Cornish Lithium, 280–2
Corpus Hermeticum, 210–11, 222
Crème Activa, 129
Crete, 67, 261
crinoids, 2
Cruquius pump, 45, 49–50
cryptocurrencies, 125
crystal healing, 174
cuckoos, 117
cuneiform, 72
Curie, Marie, 128, 130–4, 141, 148, 171
Curie, Pierre, 130–1
Cusco, 248

cyanide, 266
Cyprus, 25
Czech Geological Survey,
166–7

Dadiani family, 259
Dante, 196, 243
Danube, river, 74, 105, 217
darink-delven, 65
Dartmoor, 11
Darwin, Charles, 114, 146
David, King, 183
de Boodt, Anselmus,
162–3, 182
De Dood polder, 54–5, 63
Deason, John, 248
decontaminant plants, 286
Dee, John, 164, 175
Delft pottery, 44
della Porta, Giambattista,
160–1
Demon of Death, 70
Descartes, René, 115
Devensian ice, 56
Devon Great Consols mine,
18
diagenesis, 1
Dizdarevic, Tatjana, 198–9
DNA, 79, 144, 148
Doggerland, 57
Dolra, river, 260
Doramad Toothpaste, 128
Dordrecht, 53, 64–5
Dostoevsky, Fyodor, 94
Dürer, Albrecht, 161, 243
Dutch Golden Age, 48

earthworms, 284
East Anglia, 64
East India Company, 124
echinoids, 2
Eddystone lighthouse, 12
Egypt, ancient, 8, 20, 67–9,
91, 152–3, 208–11,
222–3, 249, 273
El Castillo Cave, 7

El Greco, 94
Elagabalus, Emperor, 153
Eldred, Julius, 214
Eliade, Mircea, 19, 208
Elizabeth I, Queen, 164
Emerald Tablet, 154–5,
222, 224
emeralds, 178
Emesa, 153
empiricism, 159, 165, 193,
222, 231
Enguri, river, 259, 265–6,
269
Enlightenment, 20, 60, 145,
160, 170, 222, 224,
231
Erzgebirge, 97, 99, 108,
110, 132–4
Eurovision Song Contest,
235

faience, 8
Fal, river, 22, 26, 40, 72,
90, 280
Fastnet Rock lighthouse, 12
Fawcus, Arnold, 230
feldspar, 10, 118
Ficino, Marsilio, 210–11
First World War, 251
Fitzgerald, Penelope, 238
Flood, the, 57, 65
fluorite, 1
Flynn, Dennis O., 98
Forbes, James, 30
Foreman, Carl, 268
Frederick the Great, 58, 60
Frisian Islands, 46
Fu Hao, 68

Gabriel, Archangel, 156,
168
Gaia hypothesis, 146–7
Galen, 195
galena, 1
Gambashidze, Irina, 254–6
garnets, 162
geodes, 1, 21, 166

Georgiana, Duchess of
Devonshire, 170
gersdorffite, 218
Gilchrist, Alexander, 244
Gilgamesh flood, 57
Giráldez, Arturo, 98
Gironde, river, 26, 72
Glastonbury Abbey, 164
glyptic mosaics, 183
Gnostics, 210, 224
Goethe, Johann Wolfgang
von, 103–5, 110,
112–19, 121–6, 160,
171, 196, 202, 211,
224, 242
Faust, 59–62, 107, 115,
124, 126, 230
'The Metamorphosis of
Plants', 122–3
*The Sorrows of Young
Werther*, 103
'Über den Granit',
118–19, 123
Veloziferische concept,
124–5, 252, 279
goethite, 171
gold, 10, 12, 14, 17, 19, 21,
24, 29, 67, 70, 94, 132,
152–4, 161, 170, 172,
174–6, 185, 188, 209,
219, 239, 246–76
currency, 247, 251
destructive industry, 266
and Nebra Sky Disc,
78–81, 87–93, 217
'potable gold', 249
prospecting, 248–9, 251
sluicing, 264, 270, 274
and transmutation, 148,
164, 194
Gold Coast, 250–1
gold standard, 251
Golden Age, 15, 48, 94
Golden Fleece, 203, 257–8,
267, 273, 277
Gorky, Maxim, 251–2
Goss Moor, 22

graphene, 25
Great Debasement, 98
Great Lakes, 214
Greece, ancient, 91, 185, 208, 210, 257
Greenland, 152
greywacke, 1, 122
Gulf of Trieste, 199
Gun-Yu flood, 57
gur, 145
Gutenberg Bible, 215
gypsum, 1

Haarlem, 45, 48
Haarlemmermeer, 44–6, 49–51
Habsburg-Lothringen, Archduke Stephan von, 171
Hájek, Tadeáš, 175
Handsteins, 183
Haraldskær Woman, 43
Harvey's of Hayle, 45–6, 49
Harz Mountains, 97, 99–100, 104–9, 112, 116–17, 120–1, 124, 171
Hatsheput, Queen, 80
Hayle, river, 10
Heaney, Seamus, 42
helium, 277
Helms, Mary, 80–1, 220–1, 223
henbane bell (*Scopolia carniolica*), 200, 202–3
Henry VIII, King, 98
Hephaestus, 69–70
Herder, Johann Gottfried, 113
hermaphrodites, 205
Hermes Trismegistus, 154–5, 168, 210–11, 222
Hermetic tradition, 113, 159, 161, 164, 183, 185, 206, 210–12, 222–4

Blake and, 229–30, 240
'Green Hermeticism', 223
see also alchemy; natural magic; Philosopher's Stone; transmutation
Herodotus, 25, 248–9
Heroic Age, 15
Hesiod, 15, 94
Himalayan singing bowls, 153
Hippocrates, 16
Hitler, Adolf, 150
Hittites, 68
Hladnikia pastinacifolia, 200
Hobson, Anthony, 229–30
Hochkönig, 220
Hodgson, Frederick, 250
Holy Roman Empire, 99, 157, 159, 204
'holy smokes', 17
Homo erectus, 6
Hořčický, Jakub, 175
Hotel Radium Palace, 134–5, 142
houndstooth calcite, 171
huayradores, 95
Humber, river, 72
Hurrians, 68
Hussein, Saddam, 58
Hutton, James, 202
Huxley, T. H., 119
hydrogen, 194, 277

ibn Hayyan, Jabir, 184–5
ice ages, 114
Idrija, 188–96, 199–204, 207, 209, 211
Ieli, 265–6
Iliad, 69
Incas, 246
India, alchemy, 20, 155, 249
Indian Ocean, 72, 80, 215
Industrial Revolution, 12, 90, 280
intermaxillary bone, 114
Inuit, 152

Iron Age, 15, 152, 209
iron, 5–8, 13–16, 19, 34, 37–8, 68, 91, 96, 139, 151, 161, 252, 257
Blake and, 224–5
bog iron, 41–3
and brain function, 8
enters food chain, 285
galvanic corrosion, 215
hatred of, 208–9
iron ore, 74, 77
from meteorites, 152–4, 247
smelting, 208
iron pyrites, 1, 5
Isleham hoard, 71
Isles of Scilly, 11, 25, 248

Jáchymov (Joachimsthal), 132–3, 134, 138, 141–2, 145, 149–50, 156, 158, 183, 200
jade, 182
Jain, Naveen, 156
James, William, 127
jasper, 1, 162, 170
Java, 69

Kalevala, 42, 91, 154
Kandinsky, Wassily, 147
kaolin, 8
kaolinite, 88
Kay, Paul, 6
Kelley, Edward, 164, 175
Kepler, Johannes, 164, 185
Kerouac, Jack, 275
Keweenaw community, 214
Keynes, John Maynard, 223
Kiesloch, 83–4
killas, 285
kinzigite, 171
Kipper- und Wipperzeit, 98
klepduiker valve, 47
Klondike, 273
Krenz, Michael, 92
Kresen Kernow, 33
Kreuning, Jippe, 50–1

Krušné Hory, 132
Kulpenberg, 93
Kunstgräben, 106–7, 117
Kvemo Kartli, 263
Kyrgyzstan mercury mine,
    199–200

Lake Superior, 214
Lamarck, Jean-Baptiste, 224
Lamb, Charles, 243
Land of Punt, 80
Land's End, 20
Langedijk, Karla, 163
lapis, 80, 166, 170
Las Médulas, 249
Lascaux Caves, 7
Laurium, 94
le Carré, John, 238
lead, 8, 17, 19, 91, 121, 138,
    142, 164, 185, 246
    atmospheric pollution,
    107–8
Leibniz, Gottfried Wilhelm,
    104–5
Lenin, V. I., 252
Lewis, C. S., 238
Lincoln, Abraham, 187
Lindow Man, 43
Linnaeus, Carl, 201–2
liroconite, 213
Lithia Springs, 278
Lithiated Lemon Soda, 278
lithium, 277–82
    'clean' lithium, 282
    medical uses, 277–8
    South American triangle,
    278, 282
lithium-ion batteries, 13,
    277, 286
llamas, 95
Lloyd George, David, 250
Locke, John, 222, 224
Loew, Rabbi, 206
London Chartered Bank of
    Australia, 248
London Radium Institute,
    142, 148

Los, 225
Lovelock, James, 146–7
lunar eclipse, 116
lung cancer, 136, 140, 144,
    150
Luther, Martin, 100

Mackenna's Gold, 267–8
magnesium, 192–3, 285
malachite, 8
Mandelstam, Osip, 94
manganese, 8
Manichaeans, 3
Mansa Musa, 250
Maraszek, Regine, 90
Maria the Jewess, 184
Maria Theresa, Empress,
    200
Marsh Arabs, 59
Marx, Karl, 124
Mathesius, Johannes, 145,
    149
Medea, 258
Medici, Cosimo de',
    210–11
Meier, Hermann, 110–11
Meier, Richard, 110–11
Meller, Harald, 87
Mendeleev, Dmitri, 127
mercury, 91, 96, 142,
    184–212, 215
    and alchemy, 184, 186
    medical uses, 186–7
    methylmercury, 187–8
    'philosophical mercury',
    204–6, 225
    poisoning, 187–8, 266
    smelting, 207–9
    and Venetian mirrors,
    204
Mercury, 91, 185–6, 205–6
Mesolithic period, 57, 247
Mesopotamia, 67, 72, 91,
    255
Mestia, 267–9
meteorites, 91, 152–3, 156,
    159, 167–8, 170, 173

Meuse, river, 40, 51, 57, 77
Mexico Towans, 10, 27
mica, 5, 118, 122
Michael, Antonín, 175
Mikhael, Archangel, 262
Minamata disease, 187–8
Minoans, 68
Mittelberg, 79, 92–3, 116
Mitterberg, 216–18, 234,
    240
Mizutani, Hayato, 92
Mojsov, Bojana, 209–12
moldavites, 167–9, 172–5,
    177–83, 231
monoclinal flexures, 1
Moon Express, 155
Morgan, J. P., 171
Morris, William, 17
Moselle, river, 83, 85
Mount Mezir, 260–1
Mount Ushba, 260
Munch, Edvard, 178
Murillo, Bartolomé
    Esteban, 94
Music of the Spheres, 154
Mussolini, Benito, 58
Mycenaeans, 68
Mystras, 211, 224, 238–9

Nancledra, 28
nanotechnology, 216
Napoleon Bonaparte, 24,
    63, 103
National Museum of
    Ireland, 42
Native Americans, 69, 214
natural magic, 159–62, 165,
    169, 182
Nature, 119
Navajo, 154
Nazis, 109–10, 149–50
Neanderthals, 6
Nebra Sky Disc, 78–9, 81,
    86–93, 116, 217, 220
nematodes, 284
Neolithic Revolution, 13
Neptunists, 202

Nero, Emperor, 17
Neva, river, 58
Newton, Isaac, 115, 155,
    222–4, 232, 251
niacin, 156
nickel, 24, 138, 154, 168,
    218
Nieuwe Merwede, 62, 66
Nigoriani, Narkis, 262–5,
    268
Nikolaj Camp, 150
nitrogen, 194
Norse mythology, 69, 154
North Sea, 48, 53, 64
Novalis, 211, 224, 238–9
Nriagu, Jerome, 17
nuclear fusion, 277
nuclear power, 133
nuclear weapons, 133,
    148–9, 158

ochre, 5–9, 13, 16, 170
Oderbruch, 58
Omphalos stone, 153
Ontonagon, river, 214
onyx, 183
opals, 170
opuka limestone, 158
Order of the Golden
    Fleece, 258
Orion constellation, 70
Osiris cult, 209
Ottersluis, 62
Ovid, 14
Oxus culture, 68

Panama Canal, 105
Paracelsus, 19, 186, 193–6,
    204, 211, 242
Parkinson's disease, 215,
    278
Paulicians, 4
Peary, Robert, E., 152
peat, 41–67, 106, 116–17,
    174
Peck, Gregory, 267–8

periodic table, 127
Perkmandic, 192, 195
Peter the Great, 58, 60
Philip II, King of Spain, 258
Philip's Planisphere, 78
Philosopher's Stone, 148,
    155, 158, 160, 164,
    166, 168, 175
Phocylides, 247
Phoenicians, 25–6
Phrygians, 154
phytoextraction, 286
Piff! Paff!! Pouf!!!, 129
Pilpani, Eka, 265
Pilpani, Vakhtang, 263–5
pirnelite, 171
pitchblende, 132–3, 138,
    140–2, 149–50, 171
Plato, 94, 210
Pleiades, 70, 78
Pliny the Elder, 46, 249
Plotinus, 210
Plutonists, 202
plutons, 11–12, 87, 280
Poldice mine, 16, 18
Polhem, Christopher, 104
polonium, 128
Pontine Marshes, 58
potassium, 285
Prague, 157–61, 164–6,
    168–9, 172–3, 175–7,
    182, 204–6, 258
Prague Jubilee Exhibition,
    173
pre-Columbian
    civilisations, 248
Příbram, 169, 171
Price Revolution, 97, 125
Principe, Lawrence, 206
prisca theologia, 212, 222
Procopius of Byzantium, 46
Prometheus, 14–15, 69,
    257
Protestant Reformation, 98
protozoa, 284
Pryce, William, 29

purpurites, 170
Pushkin, Alexander, 94

Qin, Emperor, 186
quartz, 1, 8, 88, 112, 118,
    122
quartzite, 6, 158

radiobes, 146
Radithor, 130–1, 136
radium, 17, 127–51, 174,
    188
    and atomic theory,
        146–8
    commercial products,
        128–31
    medical uses, 130
Radium Dial Company,
    131
radon, 135–8, 140, 144,
    198
Rammelsberg, 109
Ramsay, Sir William,
    141–2, 146, 148
Ramses II, pharaoh, 80
razors, bronze, 79
Red Lady of Paviland, 8
Red River, 10
Red Sea, 72, 80, 215
Red Tower of Death, 150
Reformation, 47, 98
Rembrandt, 228
Renaissance, 8, 20, 44, 98,
    154, 186, 188, 194,
    204, 210, 212, 222–4,
    229
Rhine, river, 22, 40, 45–6,
    51, 57, 72, 74, 77, 81,
    85, 105
Rijksmuseum, 52
Ringvaart, 49–50
RNA, 156
rock art, 6–7
Rodovský, Bavor, 175
Rosarium Philosophorum,
    205–6

Rosenberg family, 175–6, 182
Rosicrucians, 196
Royal Cornwall Museum, 26
Royal Mint, 251
Royal Society, 128
Rudolf II, Emperor, 157, 160–6, 169, 175–6, 182–3, 185, 204–5, 258
his lion, 164, 176
Rueff, Jacques, 124
Ruskin, John, 171, 227
Russian Revolution, 252
Rutherford, Ernest, 147–8
Ryugu asteroid, 155

St Elizabeth's Day Flood, 52–3, 64–5
St Ives, 141–4
St Ives Bay, 10
St Just-in-Penwith, 26
St Michael's Mount, 25–6
St Petersburg, 58
Sakdrisi, 254–6
Salar de Uyuni, 278–9
Saxony-Anhalt, 86–7
Scheldt, river, 40, 51
Schubert, Franz, 187
Scientific Revolution, 20, 60, 116, 159
sclerosing cholangitis, 245
scopolamine, 202–3
Scopoli, Giovanni Antonio, 200–2
Sea Beggars, 49
sea-level rise, 56–7
Seamon, David, 114
Sendivogus, Michael, 164
serpentine, 1
Shakespeare, William, 196
Shatapatha Brahmana, 249
Shaw, George Bernard, 131
Sheba, Queen of, 249
Sherrington, Charles, 113
Siberia, 43–4, 69, 101

silica, 10, 167–8, 182
silicates, 41
silver, 10, 14, 29, 88–9, 91, 94–126, 169, 174–6, 239, 246, 248, 251, 257, 279
effect on prices, 97–8, 125
Harz mines, 103–9, 112–13, 116–17, 124
Jáchymov mines, 133, 138, 140, 145, 158
model mines, 109–11
patio process, 204
silver antimony, 171
Spanish trade, 94–8, 124
Silver Age, 15, 94
silver denarii, 108
Sinop, 8
'Sipho-Machine', 104
sky, and belief systems, 154
Smale, Courtenay, 88–9
Smirnov, Vladimir, 190
Smith, Adam, 98
Smithsonian Institute, 214
snake pine, 200
Socrates, 14, 94
Soddy, Frederick, 147–8
soil, 283–9
solar trinity, 210
Solomon, King, 249
Sophocles, 94
South Crofty, 27, 280
South West Water, 282
Soviet Union, 252–3, 266, 268, 270–1, 275
speleothems, 37–8
sphagnol soap, 42
spinthariscope, 148
SS *Seestern*, 73–7, 81–6
staddle stones, 2
stalactites, 37
Stalin, Josef, 150, 253
stanene, 25
Stasi, 116
Stennack, river, 143
stephanite, 171

Stewart, Susan, 110
Stolberg, 100–1, 106
Stolberg-Stolberg, Prince Jost-Christian, 100–2
Stolberg-Stolberg, Princess Sylviane, 101–2
Stonehenge, 71
Storegga Slide, 57
Strabo, 257
Stream of Ocean, 70
Strindberg, August, 202
Suez Canal, 105
Sulawesi, 7
sulphur, 185–6, 194
Sumerians, 68, 152
sun-boats, 79
Svan language, 262
Svaneti, 257–9, 261–5, 267–9, 272, 276, 280
Svatá Hora, 169
Svatava valley, 134
Svornost mine, 138, 149
syphilis, 17, 130–1, 186–7
Syrrus, Claudius, 175

tamkaru, 72
tantalum, 24
Tanzania, 87
Tarih-i Hind-i Garbi, 96
tektites, 167–8, 170, 174, 182
Tengrism, 154
thalers, 102, 133
Thasos, 8, 249
Theophrastus, 185
Thomsen, Christian Jürgensen, 15
thorium, 147
Thoth, 210
tin, 10–13, 16, 24–40, 45, 91, 96, 132, 143, 217, 264, 280–1
alluvial, 21–2, 25, 31, 87, 288
Atlantic trade, 25–6
and bronze, 22, 25, 67, 69, 72, 80, 213

as food preservative, 24
miners' dialect, 31
and Nebra Sky Disc,
    87–9
origins of, 25
St Mawes ingot, 26
sluicing, 27–9
stanene ('2D tin'), 25
Tofana, Giulia, 17
Tollund Man, 43
Tolstoy, Leo, 94
Torgsin stores, 252–3
tourmaline, 1
transmutation, 147–8, 155,
    159, 164, 194, 204,
    222
Transylvania, 71
Tregenna Castle Hotel, 142
tria prima, 186, 194
Trianon Press, 229–30
trilobites, 170
Trump, Donald, 59
Tuaregs, 249
tungsten, 24
Tunupa, 278
Tutankhamun, pharaoh,
    153
Twain, Mark, 251, 278

Ugarit, 69, 71–2
Undark paint, 130

Úněrice, 154
uranium, 128, 138, 149–50
US Radium Company, 131
Ussher, James, 202

vajras, 153
van Beieren, Albrecht, 65
van den Vondel, Joost, 48
van Wijk, Wim, 51, 54,
    62–6
Vandkilde, Helle, 79–80
Variscan orogeny, 23
Varna tombs, 247
Velázquez, Diego de, 94
Velvet Revolution, 169
Venice, 199, 204
Venus of Willendorf, 173
Victoria, Queen, 17
Vikings, 41–2
Virgin Earth, 113
Virgin Mary, 169
Vltava, river, 158, 165,
    167–8, 175, 206
Vok, Peter, 175–6
Vulpius, Christiane, 121

War of the Golden Stool,
    250–1
Watersnoodramp, 64
de Waterwolf, 48, 51

Weimar, 103–4, 112, 121,
    125–6
Welcome Stranger, 248
Wells, H. G., 127
Westphal, Henry, 86, 90, 92
Wheal Ellen, 31–3
Wheal Jane, 27
Wheal Trenwith, 141–2
White, Gilbert, 201
Wicksteed, Joseph, 228,
    233
Windle, Sir Bertram, 146
windmills, 44–5, 49–51,
    104–5
wolves, slaughter of, 58
wristwatches, 130–1

Yan Liu, 186
Yde Girl, 43
Yucatán Peninsula, 8
Yukon, 251

Zevenbergen, 65
zinc, 10, 24, 194, 285–6
'zombie fires', 44
Zosimos of Panopolis, 184